T0296754

Artificial Intelligence for Future Generation Robotics

Artificial Intelligence for Future Generation Robotics

Edited by

RABINDRA NATH SHAW

Department of Electrical, Electronics & Communication Engineering, Galgotias University, Greater Noida, India

ANKUSH GHOSH

School of Engineering and Applied Sciences, The Neotia University, Kolkata, India

VALENTINA E. BALAS

Department and Applied Software, Aurel Vlaicu University of Arad, Arad, Romania

MONICA BIANCHINI

Department of Information Engineering and Mathematics, University of Siena, Siena, Italy

Elsevier
Radarweg 29, PO Box 211, 1000 AE Amsterdam, Netherlands
The Boulevard, Langford Lane, Kidlington, Oxford OX5 1GB, United Kingdom
50 Hampshire Street, 5th Floor, Cambridge, MA 02139, United States

British Library Cataloguing-in-Publication Data
A catalogue record for this book is available from the British Library

Library of Congress Cataloging-in-Publication Data
A catalog record for this book is available from the Library of Congress

ISBN: 978-0-323-85498-6

For Information on all Elsevier publications
visit our website at https://www.elsevier.com/books-and-journals

Publisher: Matthew Deans
Acquisitions Editor: Glyn Jones
Editorial Project Manager: Fernanda A. Oliveira
Production Project Manager: Kamesh Ramajogi
Cover Designer: Victoria Pearson

Typeset by MPS Limited, Chennai, India

Contents

List of contributors

Valentina E. Balas
Department of Automatics and Applied Software, Aurel Vlaicu University of Arad, Arad, Romania; Department and Applied Software, Aurel Vlaicu University of Arad, Arad Romania

Kanishk Barhanpurkar
Department of Computer Science and Engineering, Sambhram Institute of Technology, Bengaluru, India

Swagatam Biswas
School of Engineering and Applied Sciences, The Neotia University, Kolkata, India

Swet Chandan
Galgotias University, Greater Noida, India

Arshavee Das
School of Engineering and Applied Sciences, The Neotia University, Kolkata, India

Indrani Das
Department of Computer Science, Assam University, Silchar, India

Sanjoy Das
Department of Computer Science, Indira Gandhi National Tribal University, Regional Campus Manipur, Imphal, Manipur

Ankush Ghosh
School of Engineering and Applied Sciences, The Neotia University, Kolkata, India

Tad Gonsalves
Department of Information and Communication Sciences, Sophia University, Tokyo, Japan

Suraiya Jabin
Department of Computer Science, Faculty of Natural Sciences, Jamia Millia Islamia, New Delhi, India

Pritam Khan
Department of Electrical Engineering, Indian Institute of Technology Patna, India

Sudhir Kumar
Department of Electrical Engineering, Indian Institute of Technology Patna, India

Sampurna Mandal
School of Engineering and Applied Sciences, The Neotia University, Kolkata, India

Sarthak Mishra
Department of Computer Science, Faculty of Natural Sciences, Jamia Millia Islamia, New Delhi, India

Sk Md Basharat Mones
School of Engineering and Applied Sciences, The Neotia University, Kolkata, India

Anand Singh Rajawat
Department of Computer Science Engineering, Shri Vaishnav Vidyapeeth Vishwavidyalaya, Indore, India

Priyesh Ranjan
Department of Electrical Engineering, Indian Institute of Technology Patna, India

Romil Rawat
Department of Computer Science Engineering, Shri Vaishnav Vidyapeeth Vishwavidyalaya, Indore, India

Jyoti Shah
Galgotias University, Greater Noida, India

Pragya Sharma
G. B. Pant University of Agriculture and Technology, Pantnagar, India

Rabindra Nath Shaw
Department of Electrical, Electronics & Communication Engineering, Galgotias University, Greater Noida, India

Tarun Pratap Singh
Aligarh College of Engineering, Aligarh, India

Jaychand Upadhyay
Xavier Institute of Engineering, Mumbai, India

About the editors

Rabindra Nath Shaw is a senior member of IEEE (USA), currently holding the post of Director, International Relations, Galgotias University, India. He is an alumnus of the Applied Physics Department, University of Calcutta, India. He has more than 11 years of teaching experience in leading institutes like Motilal Nehru National Institute of Technology Allahabad, India, Jadavpur University, and others at UG and PG level. He has successfully organized more than 15 international conferences as Conference Chair, Publication Chair, and Editor. He has published more than 50 Scopus/WoS/ISI indexed research papers in international journals and conference proceedings. He is the editor of several Springer and Elsevier books. His primary area of research is optimization algorithms and machine learning techniques for power system, IoT application, renewable energy, and power electronics converters. He has also worked as a University Examination Coordinator, University MOOC's Coordinator, University Conference Coordinator, and Faculty In Charge, Centre of Excellence for Power Engineering and Clean Energy Integration.

Ankush Ghosh is presently working as an associate professor in the School of Engineering and Applied Sciences, The Neotia University, India. He has more than 15 years of experience in teaching, research, and industry. He has outstanding research experience and has published more than 80 research papers in international journals and conferences. He was a research fellow of the Advanced Technology Cell—DRDO, Govt. of India. He was awarded National Scholarship by HRD, Govt. of India. He received his PhD (Engg.) Degree from Jadavpur University in 2010. His UG and PG teaching assignments include microprocessors and microcontrollers, AI, IoT, embedded and real time systems. He has delivered invited lectures to a number of international seminar/conferences, refreshers courses, and FDPs. He has guided a large number of MTech and PhD students. He is an editorial board member of several international journals.

Valentina E. Balas is currently full professor in the Department of Automatics and Applied Software at the Faculty of Engineering, "Aurel Vlaicu" University of Arad, Romania. She holds a PhD in Applied Electronics and Telecommunications from Polytechnic University of Timisoara. Dr. Balas is the author of more than 300 research papers in refereed journals and International Conferences. Her research interests are in intelligent systems, fuzzy control, soft computing, smart sensors, information fusion, modeling and simulation. She is the Editor-in Chief of the *International Journal of Advanced Intelligence Paradigms* (IJAIP) and the *International Journal of Computational Systems Engineering* (IJCSysE), an editorial board member of several national and international journals, and is an evaluator expert for national and international projects and PhD theses.

Monica Bianchini received a Laurea cum laude in Mathematics and a PhD degree in Computer Science from the University of Florence, Italy, in 1989 and 1995, respectively. After receiving the Laurea, for 2 years she was involved in a joint project of Bull HN Italia and the Department of Mathematics (University of Florence), aimed at designing parallel software for solving differential equations. From 1992 to 1998 she was a PhD student and Postdoc Fellow with the Computer Science Department of the University of Florence. Since 1999 she has been with the University of Siena, where she is currently Associate Professor at the Information Engineering and Mathematics Department. Her main research interest is in the field of artificial intelligence and applications, and machine learning, with an emphasis on neural networks for structured data and deep learning, approximation theory, information retrieval, bioinformatics, and image processing. M. Bianchini has authored more than 70 papers and has been the editor of books and special issues on international journals in her research field. She has been a participant in many research projects focused on machine learning and pattern recognition, founded by both the Italian Ministry of Education (MIUR) and University of Siena (PAR scheme), and she has been involved in the organization of several scientific events, including the NATO Advanced Workshop on Limitations and Future Trends in Neural Computation (2001), the 8th AI*IA Conference (2002), GIRPR 2012, the 25th International Symposium on Logic Based Program Synthesis and Transformation, and the ACM International Conference on Computing Frontiers 2017. Prof. Bianchini served as Associate Editor for *IEEE Transactions on Neural Networks* (2003–2009), *Neurocomputing* (from 2002), and *International Journal of Computers in Healthcare* (from 2010). She is a permanent member of the editorial board of IJCNN, ICANN, CPR, ICPRAM, ESANN, ANNPR, and KES.

Preface

Artificial intelligence (AI) is one of the most prevalent topics in today's world. However, the application of AI we see today is just a tip of the iceberg. The AI revolution has just started to roll out. It is becoming an integral part of all modern electronic devices. Applications in automation areas like automotive, security and surveillance, augmented reality, smart homes, retail automation, and healthcare are some examples. Robotics is also rising to dominate the automated world. The future applications of AI in the robotics area are still undiscovered to most people. We are, therefore, putting an effort to write this edited book on the future applications of AI on smart robotics where several applications have been included in separate chapters. The content of the book is technical. It has tried to cover some of the most advanced application areas of robotics using AI.

This book will provide a future vision of the unexplored areas of applications of robotics using AI. The ideas to be presented in this book are backed up by original research results. The chapters provide an in-depth look with all the necessary theory and mathematical calculations. This book is perfect for researchers and developers to form an argument for what AI could achieve in the future, and those looking for new avenues and use-cases in combining AI with smart robotics and thereby providing benefits for mankind.

Rabindra Nath Shaw,
Ankush Ghosh,
Valentina E. Balas and
Monica Bianchini

CHAPTER ONE

Robotic process automation with increasing productivity and improving product quality using artificial intelligence and machine learning

Anand Singh Rajawat[1], Romil Rawat[1], Kanishk Barhanpurkar[2], Rabindra Nath Shaw[3] and Ankush Ghosh[4]

[1]Department of Computer Science Engineering, Shri Vaishnav Vidyapeeth Vishwavidyalaya, Indore, India
[2]Department of Computer Science and Engineering, Sambhram Institute of Technology, Bengaluru, India
[3]Department of Electrical, Electronics & Communication Engineering, Galgotias University, Greater Noida, India
[4]School of Engineering and Applied Sciences, The Neotia University, Kolkata, India

1.1 Introduction

The growth in artificial intelligence (AI) and robotics has contributed to expanded speculation on artificial intellectuals and startups, more prominent journal coverage about how this technology is transforming the world, and a growing surge in empirical studies into the effect these developments have on businesses, staff, and economies [1]. They also have a substantial improvement in results. In this chapter, we define the main principles, review current literature, discuss repercussions for organizational design, and explain prospects for market analysts and strategy scientists. Much work in this field has been focused on how economic development and labor markets are impacted by robotic use and the implementation of AI technology. Despite the major consequences for social welfare, this is also an important area for future study. In addition, the lack of detailed evidence on the implementation and use of AI and robotics ensures that much of the current work is not observational but depends on expert or crowd-sourced views [2]. In the future, enhanced data collection and organization would facilitate more concrete scientific research and encourage researchers to explore adjacent issues like performance

Artificial Intelligence for Future Generation Robotics.
DOI: https://doi.org/10.1016/B978-0-323-85498-6.00007-1

discrepancies and labor market impacts on various forms of robotics and technology for AI [3]. We need studies focused on data on how the effect of AI on growth, jobs, and compensation in businesses and on how artificial insight can influence economic results, such as creativity, dynamism, and inequalities, and distributional effects. Organizational analysts and policy scientists have many tools to help us consider how new developments impact our culture. We focus in particular on these issues as they are especially appropriate for operational and strategic scientists (Fig. 1.1).

The model should be capable of answering the following questions with respect to Robotic Process Automation (RPA):

- What kinds of businesses will implement robotic technology and AI?
- Do any management types or hierarchical structures exist that can be implemented very quickly?
- Are the business conditions affecting the decision on adoption?
- Is there a rise or a reduction in AI and robotics in professions, businesses, or regions?
- Are any measures of management or control capable of reducing or exacerbating the adverse effects of robotics?
- How does the implementation of AI and robotic technologies affect its producers, manufacturers upstream, and consumers in the same sector or market?
- In which situations do potential competitors contend with traditional incumbents by using machine learning (ML) or robotics? [4]
- How does the essence of work impact ML and robotics?
- What is the relative value of skills and activities required for a work modified by AI and robotics?
- How do ML and robots impact how human employees work on the job?
- What operational words replace or supplement AI and robotics for work?

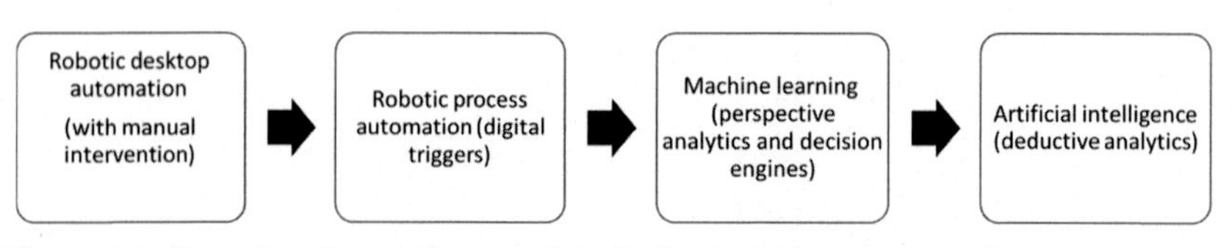

Figure 1.1 Operational work-flow model of robotic desktop automation.

Table 1.1 Comparative table for different applications and type of technique used.

Study reference	Application	Type of technique used
Asatiani et al. [5]	Robotic automation	Machine learning
Langmann and Turi [6]	Business process management	Deep learning
Huang and Vasarhelyi [7]	Auditing	Machine learning
Qiu and Xiao [8]	Cost management optimization	Machine learning
Kokina and Blanchette [9]	Accounting	Deep learning
Cernat et al. [10]	UI test automation	Machine learning

1.2 Related work

Chakraborty et al. [11] explained the most trending developments in AI which are rapidly changing the business processes. Trappey et al. [12] shows a unique intelligent patent system which provides a patent summary which will provide very necessary insights into a particular patent and help in business automation. RPA increases the overall efficiency of the business system. RPA systems are enabled with "bots" functionality, which will interact with software systems so that it will manage the human workload [3]. In the continuous process of the information systems domain, advancement in the field of Blockchain and Intelligent systems also plays an important role in development of RPA which will lead to improvement in product quality [13]. After the initial development, several steps have been taken to improve the RPA life cycle. A user-log system has been proposed which works on the knowledge of the back-office system and collects data in the form of periodic order of images and user-driven events. Additionally, RPA is used to increase productivity from the top to bottom levels of management (Table 1.1).

1.3 Proposed work

The robot systems are significantly growing in industrial automation and will increase in a growing range of situations, driven by a sustained need for expanded efficiency and better product quality [14]. Progress in

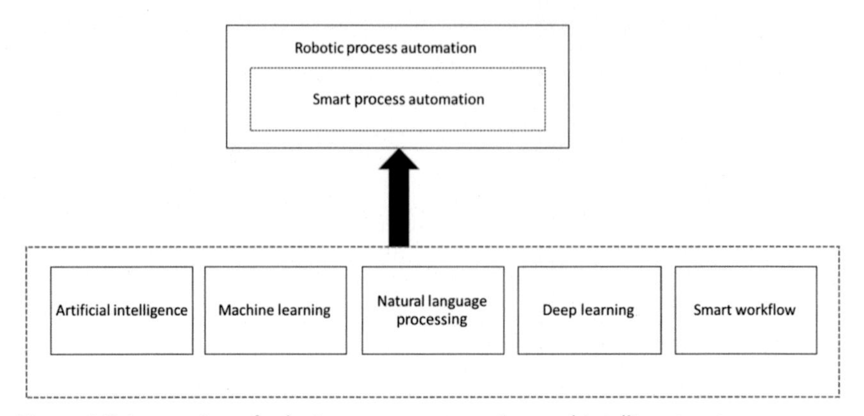

Figure 1.2 Interaction of robotic process automation and intelligent systems.

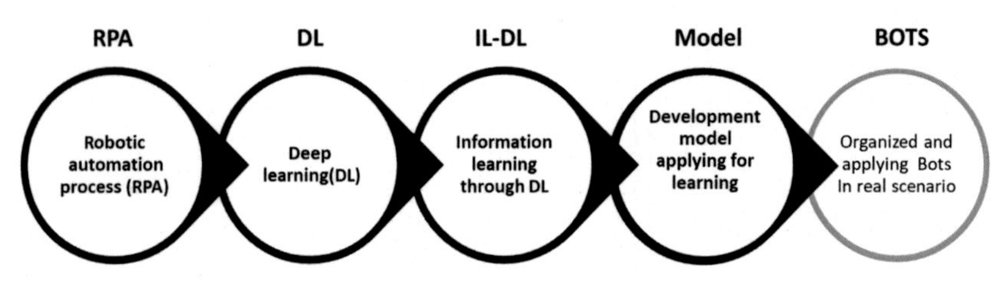

Figure 1.3 Working process of robotic automation.

depth and scope is rapidly expanding robotic systems for industrial automation. The relentless pursuit in robotic devices is usually for safer operating performance, accuracy, and reliability. To execute increasingly complicated tasks and navigate broader and diverse worlds, high intelligence and autonomy are important. This focused segment searches out state-of-the-art contributions to solve basic problems and realistic concerns in all areas of industrial automation robotic systems (William et al., 2019) (Fig. 1.2).

Industrial automation includes prototypes, preparation, and control [15]. Three sections are accepted: scheme planning and regulation, which seek to enhance robotic automation's performance; precision; and solidity [16]. The architecture of the robot was often a big problem for improving robot performance. For particular tasks, a robot that communicates with objects and environments inevitably require motion planning and reliable power. The proposed system implements a complete movement design and control system for ensuring a dual arm movement (Fig. 1.3).

Robotic process automation	Smart automation	Cognitive automation
• Update address • Represent the billing and invoicing • Represent the order entry • Prerogatives processing	• Document confirmation • Sentimental investigation • Customer segmentation • Subsequent best-action service	• Fraud detection • Market detection • Digital service agent • Recommendation engines

Figure 1.4 Key points of automation systems [17].

Yan et al. [18] suggested using a neural networks (NNs) algorithm for the optimization of algorithm prediction system parameters in a micropositioning system for the precise motion monitoring of a micropositioning system. In contrast with traditional approaches, our proposed approach, the Control Strategy, makes the precision motion monitoring of a micropositioning device clever and more flexible. By integrating visual interpretation, point cloud mechanism and awareness representation, findings indicate that the method suggested will achieve very strong output awareness of the environment. A moving target is most commonly seen in many programming situations, but for robotic manipulation, it is very difficult. Through fixing each other's faults constantly, the combined trackers will decrease their faults. On checking frames in a required database of main goal is to monitor in different locations, experimental studies reveal that, although operating separately. Robots can achieve good adaptability for performing flexible tasks and have high interactivity with people during the learning process (Fig. 1.4).

This chapter describes the assembly strategy of exploration and transfers skills between the tasks. The chapter will describe environmental dynamics with Gaussian Method during policy training in a way that decreases sampling uncertainty and increases training performance. To enhance target value estimation and to produce virtual data for transformation studies, the trained dynamic model is used. Experimental studies suggest that the suggested system increases training performance by 31% and can be applied to new activities to accelerate preparation under new policies. Simulations indicate that the suggested approach has considerable efficacy. In such a complex, dynamic system, different manipulation tasks are successfully accomplished and even in a sparse incentive environment, sample performance is increased as the research time is considerably reduced. Learning the desired human direction or motion is a crucial yet common topic in the physical contact between person and robot. It is a tailored approach to maximize human use on mundane jobs, as opposed to the unpopular notion of the "reduction in workforce." A project for process automation requires process modeling as a precondition since it serves as a plan for the project. The proposed model offers a comprehensive business process framework that allows the estimation of variables like efficiency, complexity, and full time equivalent savings from a specific process. Different criteria are taken into account, such as frequency of transition, diverse cognitive ability etc. [19]. We determine if it is appropriate for automation by using these variables and criteria. The proposed model is stronger than the standard model since it includes more parameters for individual parameters. The proposed model also analyzes the mechanism from almost every possible angle to give a more accurate detailed description, whereas the typical model's analysis is incomplete and one dimensional. Our next tasks are refining the model by taking into account further dimensions and variables, such as cost estimates, paybacks, and advantages, increasing this model to serve further market patterns, and increasing the process model's analytical base. AI-based solutions to problems relevant to robotic communication are as follows:

- First we speak about how robots understand the world and take action using Internet of Things and ML to support the ecosystem.
- More discussion will be offered on the way ML techniques are used in data collection and robotic partnerships that contribute to the development of robot communication.
- We study the key strategies and approaches to the application of ML in robotic communication for effective and reliable task performance.
- The future course of science and problems are eventually underlined.

1.4 Proposed model

A huge challenge for businesses is to understand how advantageous this reorganization is and how the modern technology notes the significance of weighing "depression" in actual expression that impairs the efficacy of face-to-face contact relative to virtual interaction. Likewise, new technology would possibly change the skills and tasks involved in many careers. New innovations alone would reform organizations and push businesses to take new developments into account. Boundaries between careers within businesses will possibly adjust as many activities are automated and people within organizations who want to utilize these technologies will definitely be more exposed to digital technologies [20]. In addition, the worker makeup will adjust to the latest and widely respected variety of skills. These improvements are often likely to be mirrored in the corporate architecture when they aim to achieve the highest benefit from their human resources. Interfirm limitations would also evolve with the increasingly common application of robotics and AI technology [21]. The robotics technologies will minimize costs dramatically within businesses, eventually resulting in fewer sales in the market [22]. Tasks which had been contracted to other entities either can be moved in-house, or corporations may find that other organizations with better access and capability to these technologies will execute more easily tasks that have previously been performed within an enterprise. Furthermore, if the technology is very unique to the organization and if the company faces the possibility of holding back an opportunistic downstream client a company may not implement emerging innovations, such as robotics. No matter how the result happens, the literature on strategy frequently shows that the current businesses cope with technical advances. Despite the obstacles faced by technological disruption, incumbents will prosper if they are "preadapted," with the ability to exploit their traditional power and assets to benefit from the emerging technologies, thus demonstrating that it would be easier for businesses to be agile and respond to modern, "smarter" robotic technologies if robotic roll users were to be in-house and with access to technical expertise. If this result is generalizable, businesses can consider using individuals with technological expertise and extending their technical knowledge facility to make the most of the possible advantages of adoption.

1.4.1 System component

Traditionally, the aerospace engineering industry is manual and robotic. The assembling of parts and their various elements is one of the most routine activities. Intensive manual labor, jigs, and special equipment are needed for this phase.

1.4.2 Effective collaboration

In the existing production methods, implementing human–robot cooperation would cut costs primarily by reducing specific equipment and times. The problem of the criteria for allocating tasks between people and robots is critical in this collaborative process [23]. The robot executes repeated

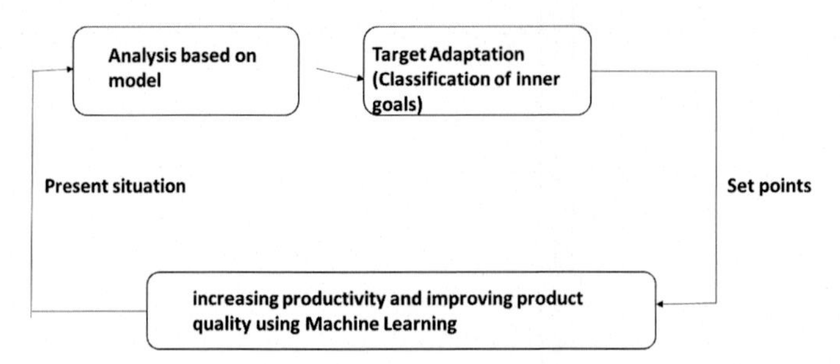

Figure. 1.5 Control of increasing productivity and improving product quality using machine learning [24].

and precise activities to maximize its operating performance, while the human being provides versatility. The worker executes a manual process that is difficult to plan, while the robot completes the process where precision is required. In order to ensure the exactness and the human–robotic relationship for secure control device coordination the solution suggested requires additional components are as follows (Fig. 1.5):

- Metrological sensors. For enhancing the precision of the robot, external sensors are used. The robot motions have high positioning discrepancies and poor direction replicability. The robot motions can be fixed to achieve more precision in positioning by incorporating a metrological device. A further argument is that the key component may be placed on the workbench with no detailed use of a metrological device.
- Safety functionalities. The robot responds similarly to human behavior. Since no obstacles remain, protection functions need to be connected to the robot device. By supplying technological details, the robot assists the worker and the worker controls the robot through the touch screen regardless of his/her skill or ability. High-level contact would allow the robot to organize behaviors with nonexpert users.

1.5 Manufacturing systems

Enhanced quality standards are necessary for the sustained market performance of goods. The consequence of the absence of certain quality

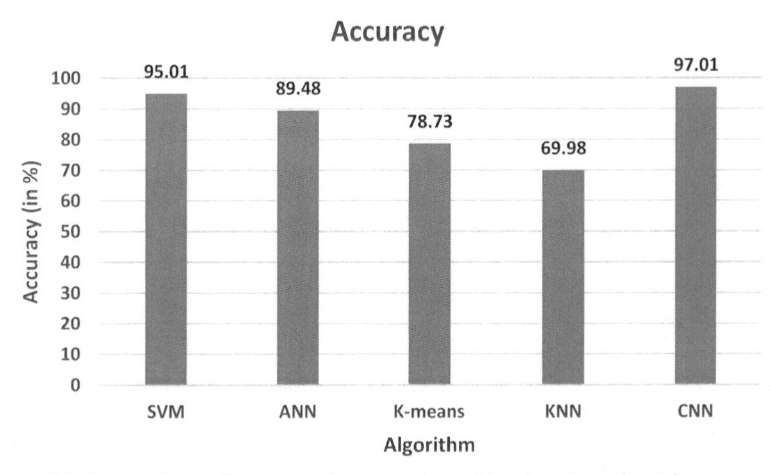

Figure 1.6 Comparison of accuracy for several machine learning algorithms.

levels will result in output flow interruptions or goods that do not fulfill the customer's requirements. For each commodity, consistency is described differently. In laser cutting, weight or surface ruggedness of the fabric are the characteristics of the fabricated part that determine the consistency. With regard to the entire phase of manufacturing, this consistency concept may be expanded to include considerations of production, for example, the usage of raw materials, the amount of time needed for production, or the workers necessary to finish the component. Manufacturing processes required to maximize certain collections of output factors are faced with the challenge of learning a great deal about the manufacturing process. Where existing systems are designed for the exact execution of setting criteria, the output consistency needs to be optimized by self-optimizing systems [25–27]. This can only be understood if there is embedded professional knowledge of the mechanism and boundary conditions which execute each step in the production chain. Examples include metal milling, cloth sewing, plate welding, injection modeling of plastic pieces, or laser radiation cutting of metal sheet. It is normal for them to process content to give it new capabilities. With every generation of machines and their control technologies, efficiency and reliability of those production systems improve. Suppliers rely on more durable materials, and where possible add faster actuators. Faster bus systems with sensor are implemented to improve circuit control and allow for quicker coordination with the unit (Figs. 1.6 and 1.7).

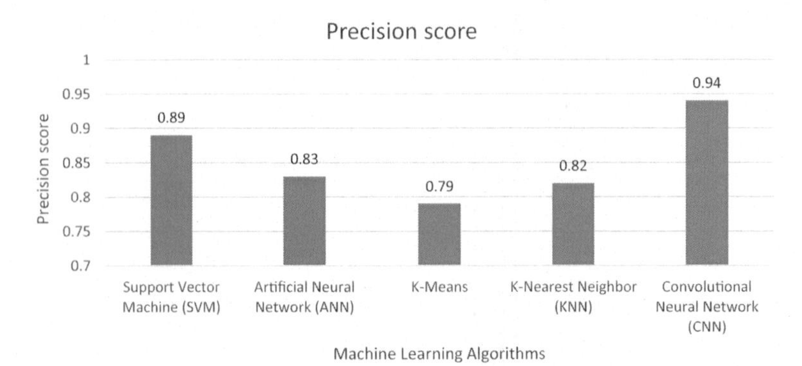

Figure 1.7 Comparison of precision score for several machine learning algorithms.

1.6 Results analysis

Results demonstrate that the symbiotic human–robot collaboration solution proposed enhances efficiency and improves safety in assembly processes for aerospace development. The strength and repeatability of the robots was successfully paired with human durability, resulting in decreased manual labor, decreased specialized equipment, and reduced time. The method suggested shows that the implementation of robotics is practical, cost-effective, and stable in the historically manual industries. With the stated modular architecture for human–robot communication, complex tasks can be semiautomated. Therefore these findings are also likely to be applied to boost the production processes of other robot-reluctant sectors. The following hardware is used for both the deployment and experiments: 12 GB RAM, i5 3.6 GHz CPU, 4 TB hard drive, and 8 GB NVIDIA graphics card. To train the NN a dataset is required. There is no dataset on user interfaces as far as our search was concerned. Thus during the creation of this research work the dataset was self-generated. In comparison, the dataset comprises three major groups: preparation (70%), validation (15%), and test (15%). Labeling is performed manually, with the assistance of our study community. Using a graphical image classification program, this role is achieved image by image. The output is shown through the xml format containing labeled synchronization and results. To represent the annotation in **J**ava**S**cript **O**bject **N**otation (JSON) format we converted the Comma-Separated Value (CSV) file into the form of script and trained it through the YOLO v3 algorithm for all label mappings. Also, Fig. 1.8 shows the analysis of the

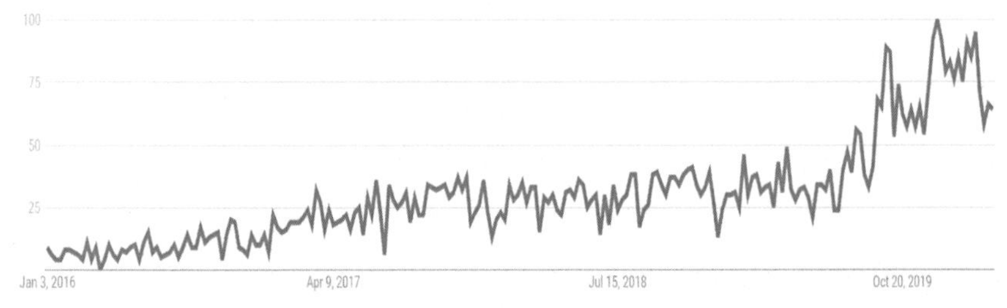

Figure 1.8 RPA analysis of duration 2016–19. *RPA*, Robotic Process Automation.

Table 1.2 Comparison of different algorithms used.

Algorithm	Accuracy	Precision score
Support vector machine	95.01	0.89
Artificial neural network	89.48	0.83
K-means	78.73	0.79
K-nearest neighbor	69.98	0.82
Convolutional neural network	97.01	0.94

number of users (in millions) of RPA over the span of 3 years. The y-axis contains the number of users and the x-axis represents the timestamp (Table 1.2).

1.7 Conclusions and future work

In manufacturing industries, robots have been used for a long time, operating securely alongside and learning from humans, but for technological and economic purposes some industries are robot-reluctant. For modern services and more complicated roles, robotics has developed and has a wider variety of capabilities than those used up to now. It is easier to incorporate peripheral technology. The architecture of the modular framework was described for the current collaborative assembly process, including power, protection, and interface modules, where people and robots will share the field of work at the same time without physical separation. Potential uses can be seen in a range of places like some robot-related sectors. In this context, the next steps in this work will concentrate on incorporating special RPA techniques as well as practical applications for other businesses, among others. The robot-reluctant sectors, such as the aerospace industry, will now incorporate cooperation between humans and robots to maximize efficiency, save resources and electricity, and boost operators' working conditions. For the existing and future manufacturing market, robots are a central feature.

References

[1] C. Di Francescomarino, R. Dijkman, U. Zdun (Eds.), Business Process Management Workshops. Lecture Notes in Business Information Processing, 2019. Available from: https://doi.org/10.1007/978-3-030-37453-2.

[2] W.M.P. Van der Aalst, M. Bichler, A. Heinzl, Robotic process automation, Bus. Inf. Syst. Eng. 60 (4) (2018) 269–272. Available from: https://doi.org/10.1007/s12599-018-0542-4.

[3] R. Syed, S. Suriadi, M. Adams, W. Bandara, S.J.J. Leemans, C. Ouyang, et al., Robotic Process Automation: contemporary themes and challenges, Comput. Ind. 115 (2020) 103162.

[4] S. Doltsinis, M. Krestenitis, Z. Doulgeri, A machine learning framework for real-time identification of successful snap-fit assemblies, IEEE Trans. Autom. Sci. Eng. (2019) 1–11. Available from: https://doi.org/10.1109/tase.2019.2932834.

[5] A. Asatiani, J.M. GarcÃa, N. Helander, A. JimÃ©nez-RamÃrez, A. Koschmider, J. Mendling, et al., [Lecture Notes in Business Information Processing] Business Process Management: Blockchain and Robotic Process Automation Forum Volume 393 (BPM 2020 Blockchain and RPA Forum, Seville, Spain, September 13–18, 2020, Proceedings) || From Robotic Process Automation toÂ Intelligent Process Automation., 2020, 10.1007/978-3-030-58779-6(Chapter 15), 215–228. Available from: https://doi.org/10.1007/978-3-030-58779-6_15.

[6] C. Langmann, D. Turi, Robotic Process Automation (RPA) - Digitalisierung und Automatisierung von Prozessen, 2020. Available from: https://doi.org/10.1007/978-3-658-28299-8.

[7] F. Huang, M.A. Vasarhelyi, Applying robotic process automation (RPA) in auditing: a framework, Int. J. Account. Inf. Syst. (2019) 100433. Available from: https://doi.org/10.1016/j.accinf.2019.100433.

[8] Y.L. Qiu, G.F. Xiao, Research on cost management optimization of financial sharing center based on RPA, Procedia Comput. Sci. 166 (2020) 115–119.

[9] J. Kokina, S. Blanchette, Early evidence of digital labor in accounting: innovation with Robotic Process Automation, Int. J. Account. Inf. Syst. 35 (2019) 100431. Available from: https://doi.org/10.1016/j.accinf.2019.100431.

[10] M. Cernat, A.N. Staicu, A. Stefanescu, Improving UI Test Automation Using Robotic Process Automation. 260–267, 2020. Available from: https://doi.org/10.5220/0009911202600267.

[11] T. Chakraborty, I. Ghosh, Real-time forecasts and risk assessment of novel coronavirus (COVID-19) cases: A data-driven analysis, Chaos, Solitons & Fractals 135 (2020) 109850.

[12] A.J.C. Trappey, et al., Intelligent compilation of patent summaries using machine learning and natural language processing techniques, Advanced Engineering Informatics 43 (2020) 101027.

[13] C. Di Ciccio, R. Gabryelczyk, L. García-Bañuelos, T. Hernaus, R. Hull, M. Indihar Štemberger, et al. (Eds.), Business Process Management: Blockchain and Central and Eastern Europe Forum. Lecture Notes in Business Information Processing, 2019. Available from: https://doi.org/10.1007/978-3-030-30429-4.

[14] M. Kirchmer, P. Franz, Value-Driven Robotic Process Automation (RPA). Business Modeling and Software Design, 2019, 31–46. Available from: https://doi.org/10.1007/978-3-030-24854-3_3.
[15] M. Majumder, L. Wisniewski, C. Diedrich, A comparison of OPC UA & semantic web languages for the purpose of industrial automation applications, in: Proceedings of the 2019 24th IEEE International Conference on Emerging Technologies and Factory Automation (ETFA), Zaragoza, Spain (2019) 1297-17-1300. Available from: https://doi.org/10.1109/ETFA.2019.8869113.
[16] E.A. Baran, O. Ayit, V.B. Santiago, S. López-Dóriga, A. Sabanovic, A self-optimizing autofocusing scheme for microscope integrated visual inspection systems, in: Proceedings of the IECON 2013 - 39th Annual Conference of the IEEE Industrial Electronics Society, Vienna (2013) 4043–4048. Available from: https://doi.org/10.1109/IECON.2013.6699783.
[17] P. Martins, F. Sa, F. Morgado, C. Cunha, Using machine learning for cognitive Robotic Process Automation (RPA), in: Proceedings of the 2020 15th Iberian Conference on Information Systems and Technologies (CISTI) (2020). Available from: https://doi.org/10.23919/cisti49556.2020.9140440.
[18] S. Yan, et al., An efficient multiscale surrogate modelling framework for composite materials considering progressive damage based on artificial neural networks, Composites Part B: Engineering 194 (2020) 108014.
[19] Y.G. Hyun, J.Y. Lee, A study on how to apply RPA (Robotics Process Automation) to improve productivity of business documents, Digital Converg. Res. 17 (9) (2019) 199–212. Available from: https://doi.org/10.14400/JDC.2019.17.9.199.
[20] F. Santos, R. Pereira, J.B. Vasconcelos, Toward robotic process automation implementation: an end-to-end perspective, Bus. Process. Manag. J. 26 (2) (2019) 405–420. Available from: https://doi.org/10.1108/BPMJ-12-2018-0380.
[21] A. Bunte et al., Why symbolic AI is a key technology for self-adaption in the context of CPPS, in: Proceedings of the 2019 24th IEEE International Conference on Emerging Technologies and Factory Automation (ETFA), Zaragoza, Spain (2019) 1701–1704. Available from: https://doi.org/10.1109/ETFA.2019.8869082.
[22] S.H. Alsamhi, O. Ma, M.S. Ansari, Convergence of machine learning and robotics communication in collaborative assembly: mobility, connectivity and future perspectives, J. Intell. Robot. Syst. (2019). Available from: https://doi.org/10.1007/s10846-019-01079-x.
[23] M. Faccio, M. Bottin, G. Rosati, Collaborative and traditional robotic assembly: a comparison model, Int. J. Adv. Manuf. Technol. (2019). Available from: https://doi.org/10.1007/s00170-018-03247.
[24] L. Pérez, S. Rodríguez-Jiménez, N. Rodríguez, R. Usamentiaga, D.F. García, L. Wang, Symbiotic human–robot collaborative approach for increased productivity and enhanced safety in the aerospace manufacturing industry, Int. J. Adv. Manuf. Technol. (2019). Available from: https://doi.org/10.1007/s00170-019-04638-6.
[25] A.S. Grema, Y. Cao, Dynamic self-optimizing control for uncertain oil reservoir waterflooding processes, in: Proceedings of the IEEE Transactions on Control Systems Technology (2019). Available from: https://doi.org/10.1109/TCST.2019.2934072.
[26] J.H. Keßler, M. Krüger, A. Trächtler, Continuous objective-based control for self-optimizing systems with changing operation modes, in: Proceedings of the 2014 European Control Conference (ECC), Strasbourg (2014) 2096–2102. Available from: https://doi.org/10.1109/ECC.2014.6862182.
[27] E. Permin, F. Bertelsmeier, M. Blum, J. Bützler, S. Haag, S. Kuz, et al., Self-optimizing production systems, Procedia CIRP 41 (2016) 417–422. Available from: https://doi.org/10.1016/j.procir.2015.12.114.

CHAPTER TWO

Inverse kinematics analysis of 7-degree of freedom welding and drilling robot using artificial intelligence techniques

Swet Chandan[1], Jyoti Shah[1], Tarun Pratap Singh[2], Rabindra Nath Shaw[3] and Ankush Ghosh[4]

[1]Galgotias University, Greater Noida, India
[2]Aligarh College of Engineering, Aligarh, India
[3]Department of Electrical, Electronics & Communication Engineering, Galgotias University, Greater Noida, India
[4]School of Engineering and Applied Sciences, The Neotia University, Kolkata, India

2.1 Introduction

Industrial robots are playing a vital role in the manufacturing industry. They are generally used as payloader, welding, painting, assembly, packaging, and other important applications. These types of robots are automated, programmable as well as can move in all three dimensional axes. The end effector of an industrial robot is one of the vital parts of the robot which is used by the robot to communicate with the environment. Therefore to place the end effector in the right position at the right time is the most challenging job for an industrial robot. The classical method to find the position of an end effector is by solving the inverse kinematics. Generally, massive numerical calculations need to be solved to get the answer for the position of the end effector. However, with the evolution of computing, solving the calculations has become easier as well as quicker. But, for an automated industrial robot, speed is a significant parameter of operation. Researchers are trying to develop faster robots to manufacture more commodities in shorter time periods. Recent developments in the field of Artificial Intelligence and machine learning allow deriving faster calculation and consequently developing faster robots. Therefore this opens a new era in robotics with the applications of Artificial Intelligence and machine learning.

Artificial Intelligence for Future Generation Robotics.
DOI: https://doi.org/10.1016/B978-0-323-85498-6.00004-6

In this chapter, we have analyzed a 7-degree of freedom (DOF) robotic manipulator with the help of D-H convention by establishing different parameters of robotic manipulators. Further, we have used Catia V5 Pro software for modeling and positioning different end-effector and manipulator positions. We have taken a number of end effectors positions (target) to calculate the joint angles. We have employed two swarm optimization: Particle Swarm Optimization (PSO) and Firefly Algorithm (FA) to find the joint angels for a given target. With the selection of random points in the workspace we have run the simulation to achieve the predetermined accuracy.

2.2 Literature review

Finding the location of the joints for the given end effector position, that is, inverse kinematics problems is a classical problem in the robotics [1,2]. As the end effector position can be written as a function of the location of the joint position, finding the inverse solution involves trigonometrical and nonlinear function, and getting the solution for a multiple DOF system becomes increasingly difficult. This problem can be solved with the help of numerical or computational methods. Classical optimization techniques were being used for solving such problems but they don't lead to the desired accuracy.

In the last 20 years many metaheuristics methods including the Cuckoo search method [3] have evolved and are being implemented for solving industrial problems. The Cuckoo search method is based on the assumption that each cuckoo will lay an egg in one nest and the best egg will be used as the new solution and rest will be discarded. This method is very helpful for providing the best solution.

In the literature [4—6] we find neural network, bee algorithm, and genetic algorithm being used to solve inverse kinematics problems. Similarly, PSO [7] and FA [8] were used to get the joint parameters for the given end effector positions.

Apart from inverse kinematics problems there are numerous other applications of robotics, such as navigation and control [9], energy minimization [10] during movement, and trajectory planning [11], where metaheuristic optimizations are being implemented to solve the problems. Recently, artificial intelligence-based optimization has also been used [12—15].

2.3 Modeling and design

For the analysis of a robot, first the workspace, link length, joint angle, etc. are to be defined. Using forward kinematics with the use of mathematical equations the end effector position is determined. As we increase the number of DOFs the mathematical complexity increases and solving inverse kinematics becomes challenging. Many algorithms are being used to solve such problems but still for higher DOF problem inverse kinematics is not a done deal.

We have started with the assigning of D-H parameters for the model. For modeling the 7-DOF robot we have employed CATIA modeling software, and using different features we have accurately shown the link and end effector position, as depicted in Figs. 2.1–2.3.

2.3.1 Fitness function

The aim of this work is to reach the desired position of the end effector from the initial input. We reduced the problem to an optimization problem with given fitness function. Essentially our approach here is to formulate the problem as an optimization problem with the target of minimizing the fitness function. We set the target to minimize the error iteratively.

$$F = \sqrt{(X_T - X)^2 + (Y_T - Y)^2 + (Z_T - Z)^2}$$

In the above expression F is chosen to be fitness function; X_T is the target position of the end effector in the x direction; Y_T is the target

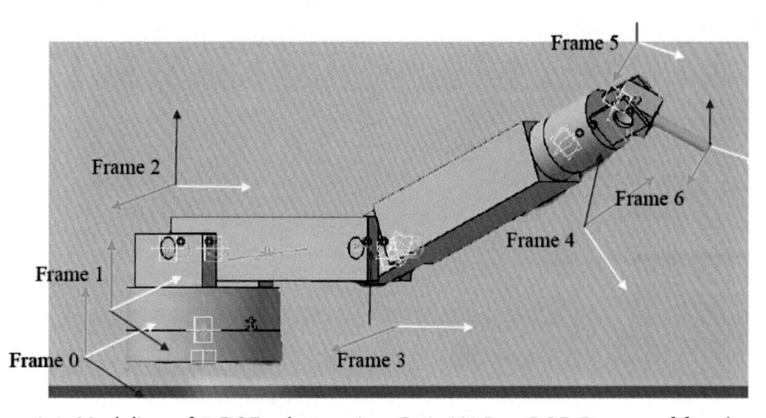

Figure 2.1 Modeling of 7-DOF robots using Catia V5 Pro. *DOF*, Degree of freedom.

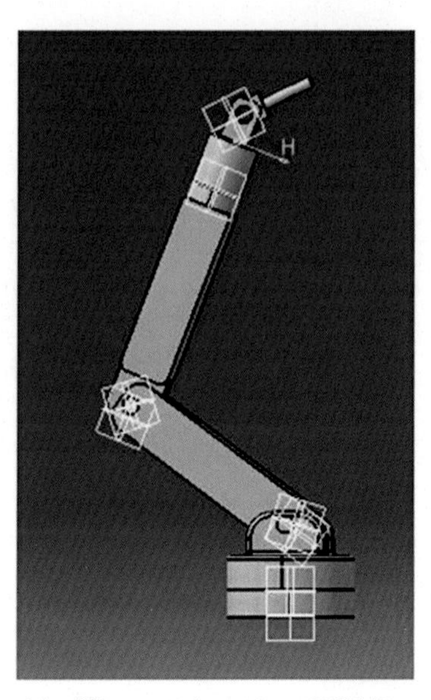

Figure 2.2 7-DOF robot with different orientation. *DOF*, Degree of freedom.

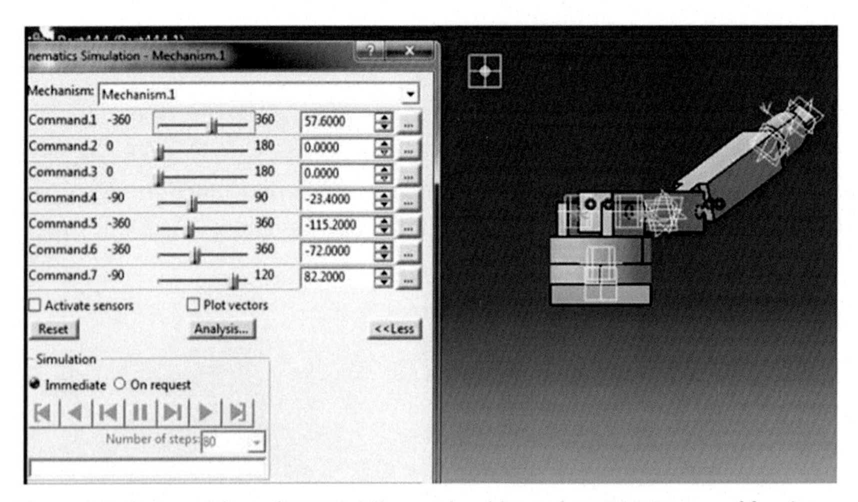

Figure 2.3 Part modeling of 7-DOF drilling and welding robot. *DOF*, Degree of freedom.

position of the end effector in the y direction; and Z_T is the target position of the end effector in the z direction. Here we have chosen the fitness function as the Euclidian distance between the target location and the present location of the end effector.

2.3.2 Particle swarm optimization

The PSO method was first proposed in 1995 by Kennedy [16]. The optimization technique is governed by personal best and global best.

In this algorithm first we define swarm size, and then we decide about the search space number of iteration etc. Further we place the position of the particle randomly and define the position too. The fitness function is calculated each time. Then the personal and global best of the fitness function is updated. Then the velocity and position of the particle is determined using the given equation each time. Finally stopping criteria is checked after running each iteration.

The updated position ($x_i(t+1)$) of the particle is a function of the last position ($x_i(t)$) of the particle and the velocity at the current location:

$$x_i(t+1) = x_i(t) + v_i(t+1)$$

And the updated velocity of the particle is given by:

$$v_{ij}(t=1) = wv_{ij}(t) + c_1 r_1 (p_{\text{best}} - x_{ij}(t)) + c_2 r_2 (g_{\text{best}} - x_{ij}(t))$$

where r_1 and r_2 are random numbers, and w is inertia weight.

The algorithm can be stated as followed. First, create a collection of the swarm on a random basis. Then evaluate the value of the fitness function for each particle, then iterate and update velocity. Further, update the position. Then, select the best solution (position) by calculating and updating the position and velocity. Finally, check accuracy and stop if the criteria are met.

The fitness function is chosen as the Eulerian square of the difference between the actual angles and calculated angles.

2.3.3 Firefly algorithm

The firefly optimization technique is another algorithm inspired by nature. Yang [3] proposed a method of optimization which mimics the behavior of fireflies. FA is being implemented nowadays to solve some of the most complex problems of mathematics, supply chains, and finance. The method is statistical and uses random variables. Fireflies of a particular type flash lights to attract another butterfly for mating, as potential prey, or other purposes.

The assumption in the fireflies algorithm is as follows: the firefly with lower intensity moves toward the brighter intensity and it decreases with the distance. The calculation of the fitness function is an indicator of the brightness.

The light intensity can be presented as:

$$I = I_0 e^{-\gamma r}$$

The measure of attractiveness can be written as:

$$\beta = \beta_0 e^{-\gamma r}$$

Movement of the firefly can be expressed as:

$$x_i = x_i + \beta_0 e^{-\gamma r^2}(x_j - x_i) + \alpha \varepsilon_i$$

where β_0 is the attractiveness when $r = 0$; no of fireflies is 50; light absorption coefficient (γ) is 0.8; attraction coefficient base value is 0.25; mutation coefficient is 0.2; mutation coefficient damping ratio is 0.997; and uniform mutation range is 0.8.

2.3.4 Proposed algorithm

In this work a new algorithm has been proposed. The new algorithm is explained in the following statements. At the beginning, D-H parameters with boundary conditions are to be defined. Then we have to start with the random solution within the workspace. We also need to define different parameters for the PSO optimization, for example, swarm size, weight, $c1$, $c2$, etc. From that calculate the fitness value for the given parameters. We also need to calculate the personal best (p_b) and global best solution (g_b). Following which, we update the velocity as well as update the current position. Finally, we compare the fitness value with the stopping criteria. If the stopping criteria matches we can exit, otherwise go to the step to calculate the fitness value for the given parameters.

The following parameters were chosen to run the PSO algorithm here in this work.

Population size = 50
Inertia weight = 2
Inertia weight damping = 0.95
Personal Learning Coefficient = 2
Global Learning Coefficient = 2.

2.4 Results and discussions

The coding was done for PSO and FA in MATLAB [17]. The initial population for the swarm was taken randomly inside the workspace of the robotic system.

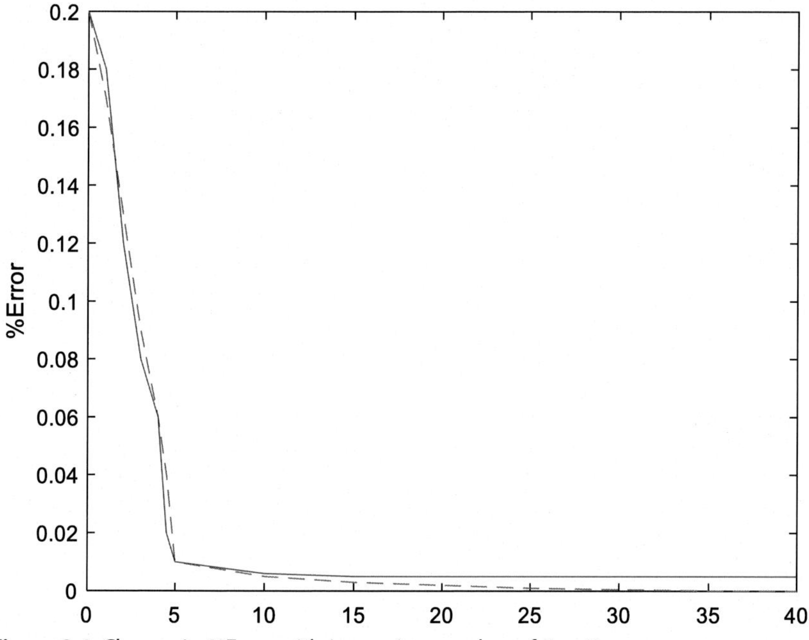

Figure 2.4 Change in %Error with increasing number of iterations

The parameters for the two algorithms were taken as per data provided in the earlier sections. The fitness function has been selected as the Euler's distance between the target position and the latest position of the end effector.

Our aim here was to minimize the error and converge to an optimal solution for each joint position. The related parameters were updated each time and checked for the convergence.

The accuracy for the simulation was selected as 10^{-4}.

The result is shown in the plot [4] generated through the MATLAB coding. From the plot it can be seen that PSO algorithm gives accuracy of over 10^{-8} comparison in just 40 iterations, whereas the FA gives an accuracy of 10^{-4} after 110 iterations Fig. 2.4.

2.5 Conclusions and future work

In the present work, we have used optimization techniques to find the joint angles of the robot for a 7-DOF robot for end effector position. We started with randomly selecting 100 points in the robotic workspace and ran the simulation in two different optimization algorithms: PSO and FA. After running the simulations, it was found that PSO outperformed FA as PSO took just 40 iterations to achieve the accuracy of 10^{-8} compared to firefly which took 110 iterations to achieve just 10^{-4} accuracy.

In future the present work can be extended for even more DOFs and some other newly developed metaheuristic algorithms can be employed to get better solution with even fewer iterations. Some real-time data can help to find the efficacy of the present method. Furthermore, two or more optimization methods can be employed and a hybrid method can be employed to get better accuracy.

References

[1] S. Kucuk, Z. Bingul, The inverse kinematics solutions of industrial robot manipulators, in: Proceedings of the IEEE International Conference on Mechatronics, ICM04 (2004) 274–279.
[2] T.P. Singh, P. Suresh, S. Chandan, Forward and inverse kinematic analysis of robotic manipulators, IRJET 4 (2) (2017) 1459–1468.
[3] X.S. Yang, Nature-Inspired Metaheuristic Algorithms, Luniver Press, (2008).
[4] Z. Bingul, H.M. Ertunc, C. Oysu, Comparison of inverse kinematics solutions using neural network for 6R robot manipulator with offset, in: Proceedings of the 2005 ICSC Congress on Computational Intelligence Methods and Applications (2005).
[5] D.T. Pham, M. Castellani, A.A. Fahmy, Learning the inverse kinematics of a robot manipulator using the bees algorithm, in: Proceedings of the 6th IEEE International Conference on Industrial Informatics, INDIN 2008 (2008) 493–498.
[6] Y. Yang, G. Peng, Y. Wang, H. Zhang, A new solution for inverse kinematics of 7-DOF manipulator based on genetic algorithm, in: Proceedings of the IEEE International Conference on Automation and Logistics (2007) 1947–1951.
[7] H. Huang, C. Chen, P. Wang, Particle swarm optimization for solving the inverse kinematics of 7-DOF robotic manipulators, in: Proceedings of the 2012 IEEE International Conference on Systems, Man, and Cybernetics (SMC), Seoul (2012) 3105–3110.
[8] S. Dereli, R. Köker, Calculation of the inverse kinematics solution of the 7-DOF redundant robot manipulator by the firefly algorithm and statistical analysis of the results in terms of speed and accuracy, in: Inverse Problems in Science and Engineering, Taylor and Francis, 2019.
[9] J. Ni, L. Wu, X. Fan, et al., Bioinspired intelligent algorithm and its applications for mobile robot control: a survey, Comput. Intell. Neurosci. 2016 (2016) 1–16.
[10] S. Kucuk, Energy minimization for 3-RRR fully planar parallel manipulator using particle swarm optimization, Mech. Mach. Theory 62 (2013) 129–149.
[11] P.V. Savsani, R.L. Jhala, Optimal motion planning for a robot arm by using artificial bee colony(ABC) algorithm, Int. J. Mod. Eng. Res. (IJMER) 2 (2012) 4434–4438.
[12] M. Kumar, V.M. Shenbagaraman, A. Ghosh, Predictive data analysis for energy management of a smart factory leading to sustainability. Book Chapter [ISBN 978-981-15-4691-4] in: M.N. Favorskaya, S. Mekhilef, R.K. Pandey, N. Singh (Eds.), Innovations in Electrical and Electronic Engineering, Springer, 2020, pp. 765–773.
[13] S. Mandal, S. Biswas, V.E. Balas, R.N. Shaw, A. Ghosh, Motion prediction for autonomous vehicles from lyft dataset using deep learning, in: Proceedings of the

2020 IEEE 5th International Conference on Computing Communication and Automation (ICCCA) 30–31 Oct (2020), 768–773. Available from: https://doi.org/10.1109/ICCCA49541.2020.9250790.

[14] Y. Belkhier, A. Achour, R.N. Shaw, Fuzzy passivity-based voltage controller strategy of grid-connected PMSG-based wind renewable energy system, in: Proceedings of the 2020 IEEE 5th International Conference on Computing Communication and Automation (ICCCA), Greater Noida, India (2020) 210–214. Available from: https://doi.org/10.1109/ICCCA49541.2020.9250838.

[15] S. Mandal, V.E. Balas, R.N. Shaw, A. Ghosh, Prediction analysis of idiopathic pulmonary fibrosis progression from OSIC dataset, in: Proceedings of the Communication Technologies (GUCON), 2–4 Oct. (2020) 861–865. Available from: https://doi.org/10.1109/GUCON48875.2020.9231239.

[16] J. Kennedy, R.C. Eberhart, Particle swarm optimization, in: Proceedings of IEEE International Conference on Neural Networks, November (1995) 1942–1948.

[17] www.mathworks.com.

CHAPTER THREE

Vibration-based diagnosis of defect embedded in inner raceway of ball bearing using 1D convolutional neural network

Pragya Sharma[1], Swet Chandan[2], Rabindra Nath Shaw[3] and Ankush Ghosh[4]
[1]G. B. Pant University of Agriculture and Technology, Pantnagar, India
[2]Galgotias University, Greater Noida, India
[3]Department of Electrical, Electronics & Communication Engineering, Galgotias University, Greater Noida, India
[4]School of Engineering and Applied Sciences, The Neotia University, Kolkata, India

3.1 Introduction

Traditionally, for bearing fault diagnosis different methods are utilized for the extraction of bearing features and then for classification of the faults embedded in a ball bearing. Features were manually extracted and separate methods were used for the classification. In the last 2–3 years, deep learning techniques have picked up all consideration for the determination of faults installed in ball bearing [1,2]. In previous work, deep learning techniques were just utilized for the order of flaws implanted in any metal roller, and various strategies for signal investigation, for example, fast Fourier transform, wavelet transform, etc., were applied separately for feature extraction of the ball bearing, and then only the neural networks were trained and tested. In any case, presently from the most recent couple of years, deep learning strategies have been utilized for both, first for the extraction of features by vibration data, and for the classification of faults of the ball bearing [3–6]. In the literature, we find many variants of deep learning approaches, for example, Recurrent Neural Network, Generative Adversarial Network, Deep Belief Network, and Convolutional Neural Network (CNN).

A variant of deep learning approaches, the 1D CNN algorithm is applied for recognizable proof and characterization of the deficiencies

Artificial Intelligence for Future Generation Robotics.
DOI: https://doi.org/10.1016/B978-0-323-85498-6.00011-3

implanted at the bearing internal raceway. The similar methodology of 1D CNN is applied for both extraction of features and for the classification of fault in a ball bearing.

3.2 2D CNN—a brief introduction

There is a neural-organic model of creature visual cortices [7] by which the CNN is biologically motivated and CNN is a deep learning algorithm based on an Artificial Neural Network. From the outset the convolution measure is utilized for picture handling and picture acknowledgment progressively; first, it is utilized for straightforward features, such as edge and corner, and afterward it is utilized for extricating complex features.

There are three main stages in the convolution operation, two stages of which are for filtering and the third stage is for classification. The architecture or the structure of 2D CNN, as shown in Fig. 3.1, consists one layer of convolutional and the other layer of pooling. To get deep in the network this combination of both layers is repeated in the model. The convolutional and pooling layer is the filter stage of the model. The input in the process is a 2D image of a ball bearing with the fault of any type. The convolutional layer generates a new feature map image from the raw input image. The feature map image includes unique features of the input image by processing the 2D data. This layer contains filters which are called convolution filters. The other filtering stage is a pooling layer which decreases the output image size of convolution filters. Then the output of these two filtering stages (hidden layers) is transferred to the third stage of one or some fully connected layers; the next output from

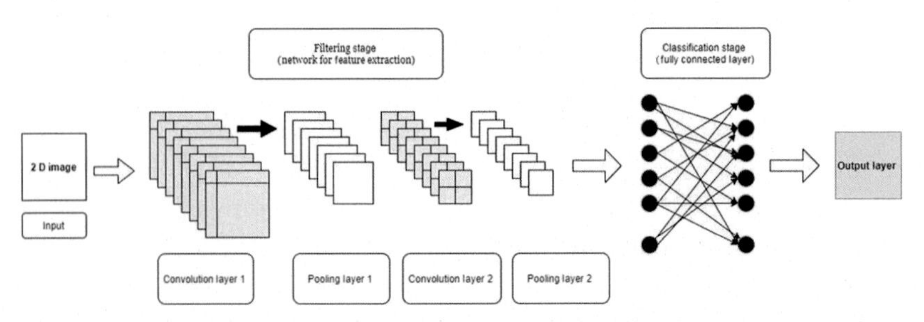

Figure 3.1 Architecture of 2D CNN first point. *CNN*, Convolutional neural network.

these layers is forwarded to the best classifier to get the detail if the fault is present in the ball bearing. These classifiers are based on SoftMax or Sigmoid functions.

In the literature there are extensive uses of profound learning approaches for fault finding [2,8]. Be that as it may, an enormous named dataset is needed to work with these methodologies. Testing and training of the algorithm are done using the dataset. The computational multifaceted nature is likewise high for these methodologies. To conquer the disadvantages of the high-level methodology of CNN, one-dimensional (1D) CNN is proposed with the upside of less multifaceted nature in calculation as 1D information is prepared in CNN layers.

3.3 1D convolutional neural network

1D CNN is the advanced methodology of conventional CNN. 1D CNN works on 1D vibration signals. Similar to the conventional 2D CNN architecture, this proposed method also consists of two layers. The principal layer is known as the convolutional layer. *B*oth the process of 1D convolution and subsampling happens in first layer. The subsequent layer is the Multi-Layer Perceptron (MLP) layer, which is indistinguishable from the completely associated layer in the regular CNN strategy. Fig. 3.2 shows the architecture or structure of 1D CNN.

1D CNN is successful for both extraction of features or highlights and the classification of defects or fault embedded in the bearing. Because of this versatile and adaptable design, quite a few layers can be taken practically for hidden layers, and then the subsampling variable or factor of the yield or output CNN layer adaptively selects the number of MLP layers and automatically decides the feature map dimension. As mentioned in Ref. [9], the process of feature extraction and classification of fault of rolling element bearings are integrated into a single 1D CNN. During the period of training of the CNN model, to minimize the error and to maximize the performance of the classification layer, the backpropagation (BP) algorithm is optimized using a gradient descent optimization approach. In the input layer of 1D CNN, there is no necessity for the manual treatment of features or information. 1D CNN layers adjust the information and the feature extraction and subinspecting or subsampling by kernel size, and the feature map is performed by shrouded neurons of the

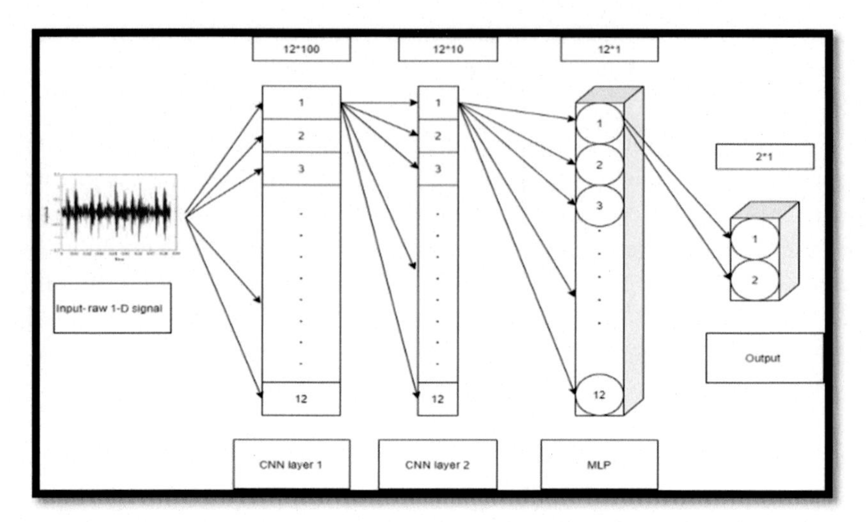

Figure 3.2 1D CNN sample consisting two CNN layers and one MLP layer. *CNN*, Convolutional neural network; *MLP*, Multi-Layer Perceptron.

convolution layer. The MLP layer is utilized for classification of the bearing fault. The output from the CNN layer is moved to the MLP layer where the 1D convolution of the 1D signal, along with kernel filter, is executed. Further subtleties including the detailing of the classifier of 1D CNN depending on the BP calculation can be found in Ref. [9].

In 1D CNN [10], as shown in Fig. 3.3, first, the weights are assigned to the neurons, and bias is defined for the CNN layers; afterward, forward propagation is utilized as the yield from the CNN layer, where ($l-1$) is the contribution to the subsequent shrouded CNN layer l.

$$x_k^l = b_k^l + \sum_{i=1}^{N_{l-1}} \text{conv1D}(w_{ik}^{l-1}, S_i^{l-1})$$

where x_k^l is input or contribution for layer l from the $l-1$ layer; b_k^l is bias of kth neuron of $l-1$ layer; w_{ik}^{l-1} is kernel from ith neuron of $l-1$ layer to the kth neuron at l layer; and S_i^{l-1} is yield of i th neuron at $l-1$ layer.

As shown in Fig. 3.2, the yield or output y_k^l is obtained from the input x_k^l of layer 1,

$$y_k^l = f(x_k^l) \text{ and } s_k^l = y_k^l \downarrow \text{ss}$$

where s_k^l is the after effect of the neuron and is afterward tested with $\downarrow$ss activity with the ss factor.

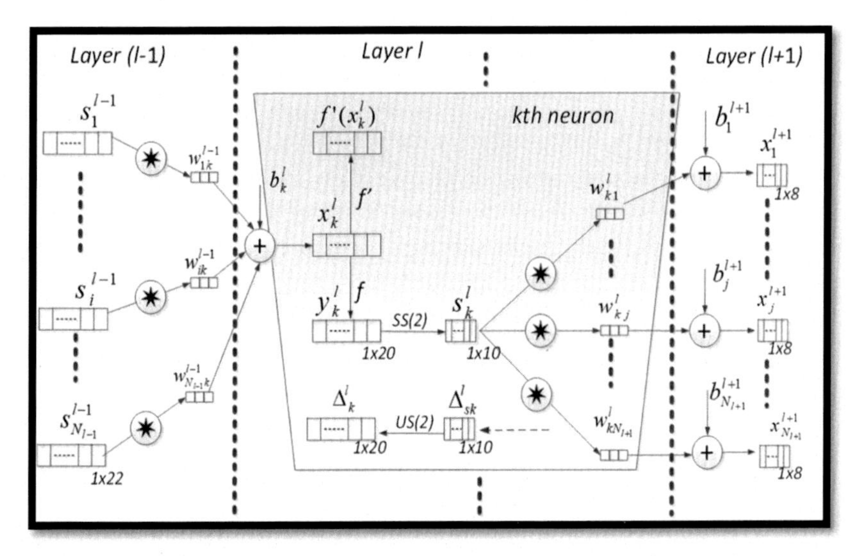

Figure 3.3 1D CNN hidden layers [10]. *CNN,* Convolutional neural network.

For an error or mistake, BP starts from the yield of the MLP layer. Let $l = 1$ show the input or info layer and $l = \mathrm{L}$ is the yield or output layer. Presently, in the information base, the quantity of classes can be considered as N_n.

The info vector is p, the objective vector of the input is t_i^p, and yield vectors are $[y_1^L, \ldots\ldots, y_{N_L}^L]$. The mean-squared error (MSE) in the layer of yield, E_p for input p is characterized as:

$$E_p = \mathrm{MSE}\left(t_i^p, \left[y_1^L, \ldots.., y_{N_l}^L\right]\right) \sum_{i=1}^{N_L} (y_i^L - t_i^p)^2$$

There are a few errors or mistakes in the model because of organization boundaries. The fundamental focal point of BP is to limit the commitment of boundaries of the organization in mistakes. For the minimization of mistakes, the subsidiary of MSE is figured concerning the individual relegated weight, which is associated with that specific neuron, k. The gradient descent method is applied to limit this commitment of organization boundaries in mistakes. By utilizing the chain rule of subordinate, the bias and weight of the neurons can be refreshed as shown:

$$\frac{\partial E}{\partial w_{ik}^{l-1}} = \Delta_k^l y_i^{l-1} \text{ and } \frac{\partial E}{\partial b_k^l} = \Delta_k^l$$

Further mathematical modeling and the BP algorithm are given in the literature [9–11].

3.4 Statistical parameters for feature extraction

In earlier work, these different features were extracted manually or by any other method in which high expertise was required, and then a different method was used for the classification of the fault in a ball bearing. But nowadays, with this advanced approach of CNN based on deep learning, features are extracted automatically, and then the required feature for further analysis is adaptively selected by the model itself based on the effectiveness and importance of the feature. In Table 3.1, we have presented different features extracted in the literature.

Table 3.1 Statistical parameters for feature extraction [12].

• Feature	• Definition
Mean value	$\overline{x} = \frac{1}{n}\sum_{i=1}^{n} x_i$
Root mean square (RMS)	$\mathrm{RMS} = \left[\frac{1}{n}\left(\sum_{i=1}^{n} x_i\right)\right]^{1/2}$
Standard deviation	$\sigma^2 = \frac{1}{n-1}\sum_{i=1}^{n}(x_i - \overline{x})^2$
Kurtosis value (KV)	$\mathrm{KV} = \frac{1}{n}\sum_{i=1}^{n}\left(\frac{x_i-\overline{x}}{\sigma}\right)^4$
Crest factor (CF)	$\mathrm{CF} = \frac{\max(\lvert x_i \rvert)}{\left[\frac{1}{n}\left(\sum_{i=1}^{n} x_i\right)\right]^{1/2}}$
Inner race ball pass frequency (BPFI)	$\mathrm{BPFI} = f_s \frac{N}{2}\left(1 + \frac{D_B}{D_P}\cos\alpha\right)$
Outer race ball pass frequency (BPFO)	$\mathrm{BPFO} = f_s \frac{N}{2}\left(1 - \frac{D_B}{D_P}\cos\alpha\right)$
Ball spin frequency (BSF)	$\mathrm{BSF} = f_s \frac{N}{2D_B}\left(1 - \frac{D_B^2}{D_P^2}\cos\alpha\right)$
Cage frequency or fundamental train frequency (FTF)	$\mathrm{FTF} = \frac{N}{2}\left(1 - \frac{D_B}{D_P}\cos\alpha\right)$

where x is vibration signal; n is number of sampling points; f_s is shaft speed; N is number of balls; D_B is ball diameter; D_P is pitch diameter; and α is contact angle between the inner race and outer race.

3.5 Dataset used

The Machinery Failure Prevention Technology (MFPT) dataset [13] is the openly accessible information for ball bearing flaw finding that is used in this work. For the MFPT dataset, a sort of NICE bearing is utilized in the test rig.

Parameters of ball bearing:
Diameter of ball: 0.23
Pitch diameter: 1.24
Number of balls: 8
Contact angle: 0

The dataset of MFPT is arranged into three segments of bearing vibration data: baseline condition, inner raceway flaw condition, and outer raceway fault conditions. For the pattern set of information, three records are tested for the sample rate of 97,656 Hz for 6 seconds for every document. For the external race flaw condition dataset, seven documents are tested for the sample rate of 48,828 Hz for 6 seconds for each record. Furthermore, for the inward race deficiency condition dataset, seven documents are examined with the sample rate of 48,828 Hz for 3 seconds for every record. All the data in the data were obtained at different load conditions.

3.6 Results

When the ball element of the bearing makes any contact with the fault or defect or flaw at any raceway of the bearing, the implication of hitting the fault will change the corresponding frequency, that is, inner race ball pass frequency (BPFI), outer race ball pass frequency (BPFO), ball spin frequency (BSF), fundamental train frequency (FTF). The raw 1D vibration signal is obtained by an accelerometer in the test rig, and is further analyzed in the 1D CNN layer. Seventy percent of the dataset is utilized for preparing the algorithm and the remaining 30% information of the dataset is utilized for testing the algorithm. Fig. 3.4 shows the crude vibration signal of the condition when the fault is embedded at the inner race of the ball bearing.

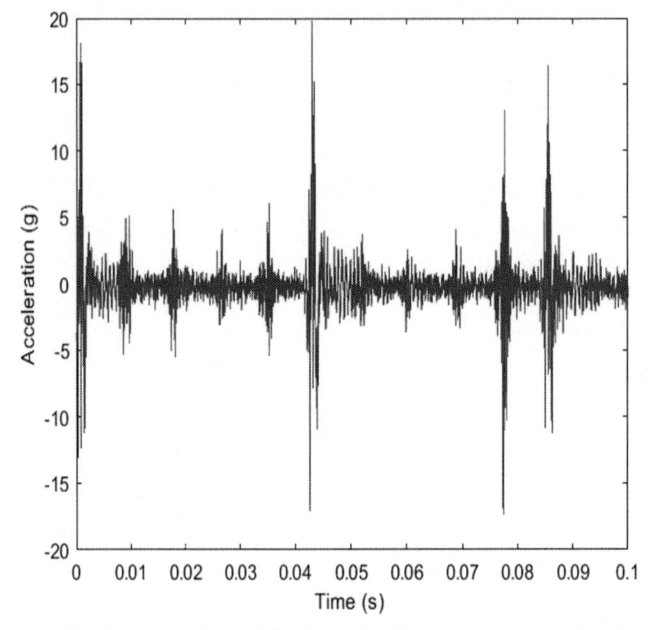

Figure 3.4 Crude vibration sign of fault at the inner raceway of bearing of MFPT dataset.

This crude information of the vibration is changed over in the frequency domain or recurrence area, which is anything but difficult to measure in the CNN layer for highlight or feature extraction. In this work, we have mainly focused on performing our analysis for a fault embedded at the bearing inner race. To get a clear visualization of raw data, the vibration information is changed over into the recurrence area or frequency domain, as shown in Fig. 3.5. The sign is then wrapped, as in Fig. 3.6, which depicts the pinnacle modification, through eliminating the noise using the sign. The CNN layer separates the component by convolution, followed by subexamining. What's more, it adaptively chooses the element as per which class will be chosen. The envelope range of a typical bearing is given to determine the distinction in signs of vibration.

With the help of signals in the time domain, kurtosis is calculated for bearing vibration data. For a random variable, its fourth standardized moment is termed as kurtosis. With the assistance of kurtosis, the impulsiveness or rashness of the sign can be estimated or the substantialness of the tail of the arbitrary variable can likewise be classified. As can be seen in Fig. 3.7, the inner raceway kurtosis is not the same as the ordinary or

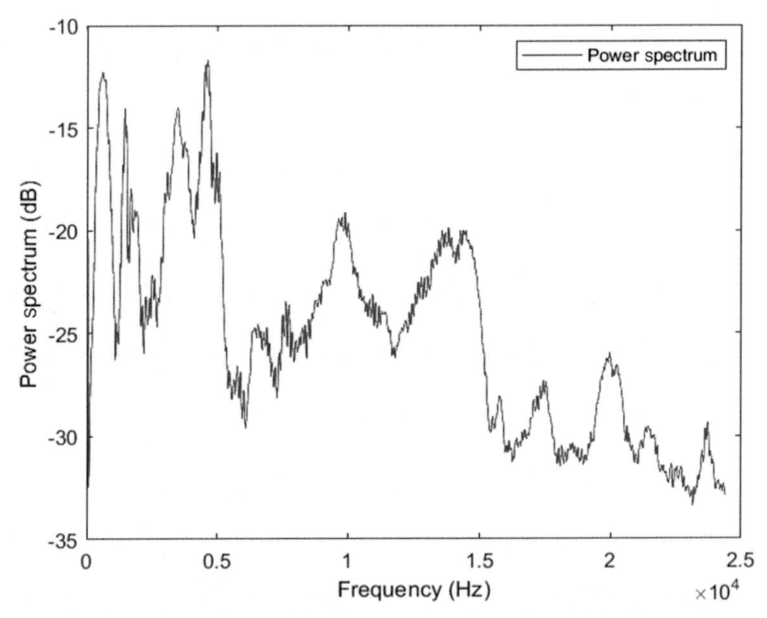

Figure 3.5 Conversion of crude vibration signal of inner raceway fault in frequency domain from the time domain.

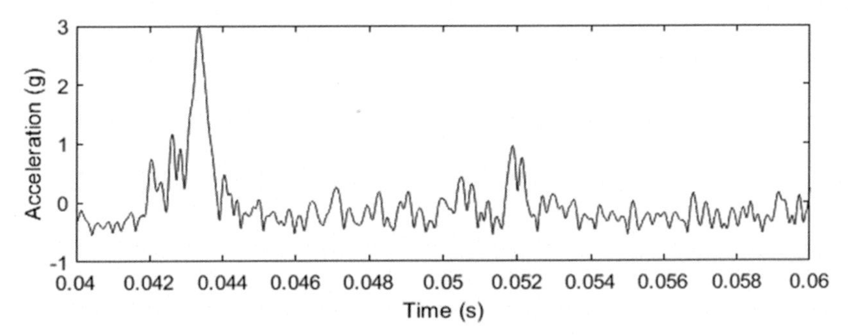

Figure 3.6 Envelope signal.

healthy bearing. Fig. 3.7 arranges the sign picture of kurtosis for the external raceway issue of the bearing, the solid or ordinary bearing, and the inward raceway issue bearing.

To validate the classifier, the testing is done with 30% labeled data of the dataset and the log ratios of the amplitude of BPFI and BPFO correctly describes the accuracy of the system. The effect of the arrangement frameworks is discovered by utilizing evaluation matrices. The outcome

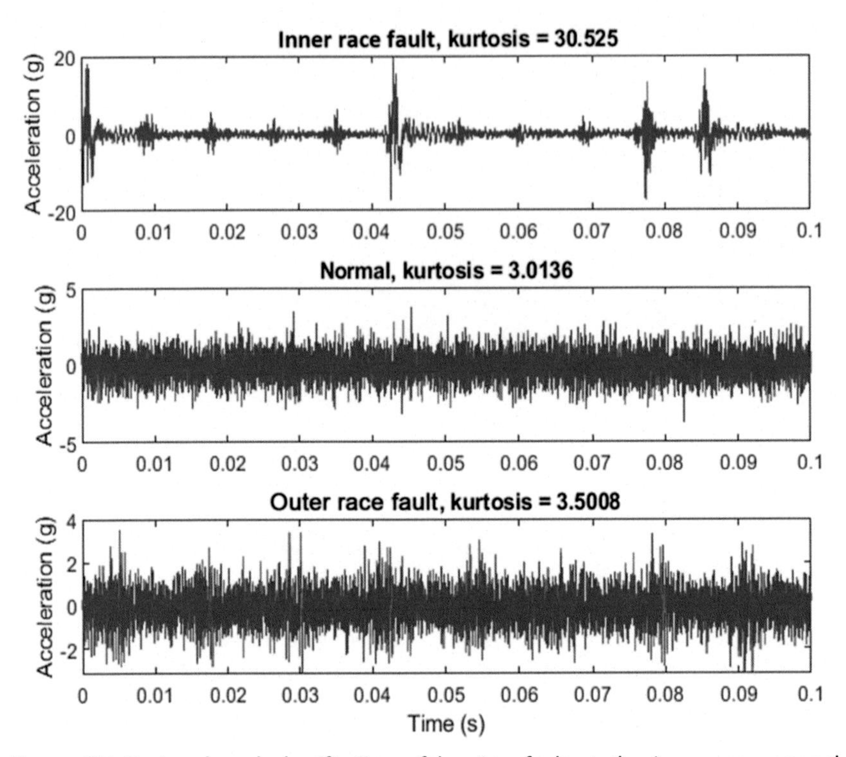

Figure 3.7 Feature-based classification of bearing fault at the inner race, normal bearing, and fault at outer race.

from the information can be introduced with regard to evaluation matrices. By and large, flawlessness and precision can be characterized in two classes, one is healthy and the other class is faulty. For any machine component with explicit matrices, classes can be referenced as true positive [TP], true negative [TN], false positive [FP], and false negative [FN]. These networks' results straightforwardly mirror the effect on state or condition.

The evaluation matrices can be processed as:

		Result
Accuracy	$\text{Acc} = \frac{[\text{TP}] + [\text{TN}]}{[\text{TP}] + [\text{TN}] + [\text{FP}] + [\text{FN}]}$	99%
Sensitivity	$\text{Sen} = \frac{[\text{TP}]}{[\text{TP}] + [\text{FN}]}$	96.1%
Specificity	$\text{Spe} = \frac{[\text{TN}]}{[\text{TN}] + [\text{FP}]}$	98.5%
Positive predictivity	$\text{Ppr} = \frac{[\text{TP}]}{[\text{TP}] + [\text{FP}]}$	97.5%

This methodology of 1D CNN achieved the targets, as the methodology provides results with a high degree of precision when contrasted with the customary strategies. The computational intricate nature of 1D CNN is less for condition monitoring for real-time systems. The purpose behind low computational intricacy is to utilize the 1D arrays, rather than 2D arrays, for the recognition of defects or flaws or faults. There is additionally a low time delay in fault or flaw distinguishing.

3.7 Conclusion

In this work, we have begun with crude vibration signal information collected from an openly accessible MFPT dataset. A deep learning technique-based engineering was planned utilizing convolutional layers, subsampling layers, and MLP layers. The neural network was prepared utilizing 70% of a named dataset and the remaining 30% of the dataset was utilized for testing of the model. Finally, test sets were simulated by using this proposed 1D CNN model. After testing, these results were analyzed and compared with the results from the literature. We further suggest that this strategy can be applied to analyze the fault in ball bearings with high precision and it can be implemented for the constant condition monitoring of bearings.

References

[1] H. Shao, H. Jiang, H. Zhang, W. Duan, T. Liang, S. Wu, Rolling bearing fault feature learning using improved convolutional deep belief network with compressed sensing, Mech. Syst. Signal. Process. 100 (2018) 743−765. Available from: https://doi.org/10.1016/j.ymssp.2017.08.002.

[2] O. Janssens, V. Slavkovikj, B. Vervisch, K. Stockman, M. Loccufier, S. Verstockt, et al., Convolutional neural network based fault detection for rotating machinery, J. Sound Vib. 377 (2016) 331−345. Available from: https://doi.org/10.1016/j.jsv.2016.05.027.

[3] P. Tamilselvan, P. Wang, Failure diagnosis using deep belief learning based health state classification, Reliab. Eng. Syst. Saf. 115 (2013) 124−135. Available from: https://doi.org/10.1016/j.ress.2013.02.022.

[4] C. Li, R.V. Sánchez, G. Zurita, M. Cerrada, D. Cabrera, Fault diagnosis for rotating machinery using vibration measurement deep statistical feature learning, Sensors 16 (2016). Available from: https://doi.org/10.3390/s16060895. Switzerland.

[5] R. Zhao, R. Yan, J. Wang, K. Mao, Learning to monitor machine health with convolutional bi-directional LSTM networks, Sensors 17 (2017). Available from: https://doi.org/10.3390/s17020273. Switzerland.

[6] X. Guo, L. Chen, C. Shen, Hierarchical adaptive deep convolution neural network and its application to bearing fault diagnosis, Meas. J. Int. Meas. Confed. 93 (2016) 490−502. Available from: https://doi.org/10.1016/j.measurement.2016.07.054.

[7] K. Fukushima, Neocognitron: a self-organizing neural network model for a mechanism of pattern recognition unaffected by shift in position, Biolog. Cybern. 36 (1980) 193–202.
[8] S. Guo, T. Yang, W. Gao, C. Zhang, A novel fault diagnosis method for rotating machinery based on a convolutional neural network, Sensors 18 (2018) 1429.
[9] S. Kiranyaz, T. Ince, M. Gabbouj, Real-time patient-specific ECG classification by 1D convolutional neural networks, IEEE Trans. Biomed. Eng. 63 (2015) 664–674.
[10] S. Kiranyaz, T. Ince, M. Gabbouj, R. Hamila, Convolutional neural networks for patient-specific ECG classification, in: Proceedings of the 37th IEEE Engineering in Medicine and Biology Society Conference (EMBC'15). Milano, Italy, Aug. (2015).
[11] O. Abdeljaber, O. Avci, S. Kiranyaz, M. Gabbouj, D.J. Inman, Real-time vibration-based structural damage detection using one-dimensional convolutional neural networks, J. Sound. Vib. 388 (2017) 154–170.
[12] M. Sohaib, C.-H. Kim, J.-M. Kim, A hybrid feature model and deep-learning-based bearing fault diagnosis, Sensors 17 (2017) 2876. Available from: https://doi.org/10.3390/s17122876.
[13] E. Bechhoefer, Condition based maintenance fault database for testing diagnostics and prognostic algorithms. <https://mfpt.org/fault-data-sets/>, 2013.

CHAPTER FOUR

Single shot detection for detecting real-time flying objects for unmanned aerial vehicle

Sampurna Mandal[1], Sk Md Basharat Mones[1], Arshavee Das[1], Valentina E. Balas[2], Rabindra Nath Shaw[3] and Ankush Ghosh[1]

[1]School of Engineering and Applied Sciences, The Neotia University, Kolkata, India
[2]Department and Applied Software, Aurel Vlaicu University of Arad, Arad Romania
[3]Department of Electrical, Electronics & Communication Engineering, Galgotias University, Greater Noida, India

4.1 Introduction

This world is headed toward a future where the skies are not occupied with only birds, manually driven airplanes, helicopters, and fighter planes. In the near future the sky is going to be occupied by unmanned vehicles ranging from unmanned drones to large Unmanned Aerial Vehicles (UAV). Some of the UAVs will be able to communicate among themselves without human interruption. It is extremely important to make commercial as well as military UAVs economical, safe, and reliable. One of the solutions is to use inexpensive sensors and camera devices for collision avoidance.

The latest automotive technologies have given solutions to these problems, for example, nowadays commercial products [1,2] are designed to sense the obstacles and avoid collision with other cars, pedestrians, cyclists, and other movable and immovable objects. Modern technology has already made progress on navigation and accurate position estimation with either single or multiple cameras [3–9], although there is comparatively little improvement in visual guided collision avoidance technique [10]. Basically, it is quite difficult to apply the same algorithm for pedestrian or automobile detection just by simply extending and modifying. Some challenges for flying object detection are given below:

- The object motion is more complex due to the 3D environment.
- There are various kinds of flying objects with diverse shapes. These flying objects can appear both on the ground and in the sky.

Artificial Intelligence for Future Generation Robotics.
DOI: https://doi.org/10.1016/B978-0-323-85498-6.00005-8

Because of that the background also becomes more diverse and complex.

- The UAV must detect potentially dangerous objects from a safe distance, where speed is also involved.

Fig. 4.1 shows some examples where flying objects are being detected on just a single image where even the human eye cannot clearly see them. When the sequence of frames is observed, these objects popped up and therefore were detected easily, which leads to the conclusion that motion cues play a very crucial role in detection.

But these motion cues become very difficult to target if the images are taken from a moving camera with featured backgrounds. It becomes a matter of challenge to stabilize as they are changing rapidly and are nonplanar. Moreover, since there are other moving objects, as shown in the first row of Fig. 4.1, there is a person in motion. But only motion is not sufficient, and the corresponding appearance must be taken care of.

In this chapter, our target is to detect and classify the object of interest present in the image or frame. The work for flying object detection in this chapter is divided into two steps:

- At first, the images of flying objects taken from various sources with different shapes and diverse background are detected and classified

Figure 4.1 Object detection sample image.

with labeled bounding box along with a confidence score predicted by the network.
- Then, the same network is applied to video to test the performance of the network while the motion is included along with the constantly changing background.

The primary motivation behind this work is to test the Single Shot Detection (SSD) algorithm to detect and classify flying objects such as airplanes and birds. The secondary objective for working on this chapter is to demonstrate how this algorithm is successfully applied for real-time flying object detection and classification for frames taken from a video feed during UAV operation.

4.2 Related work

There are three main categories in which the moving objects can be classified:
- Appearance-based methods, which rely on the appearance in individual frames;
- Motion-based methods, which rely on the information about motion across frames; and
- Hybrid methods, which combine the above two processes.

Let us briefly discuss the review of the incorporated processes:

4.2.1 Appearance-based methods

These types of methods rely on ML (machine learning) algorithms and the performances of these algorithms are quite good even in cluttered background or in complex light variation. They are typically based on Deformable Part Models [11], Convolutional Neural Networks (CNN) [12], or Random Forests [13]. Among these algorithms the best is Aggregate Channel Features [14].

These algorithms perform better for those kind of target objects where the objects are clearly visible in individual images. This is not the case applied for us. As an example, let's say, there is fog or smoke in the environment. The algorithm should be able to detect a target which is a flying object even in a hazy or foggy environment. Another example is shown in Fig. 4.1, the object is quite small and is not visible with the human eye. It was only possible to detect that object because of the motion cues.

4.2.2 Motion-based methods

These types of methods can be divided further into two subsets: The first one comprises the targets that rely upon background subtraction [15–18] and this subclass determines the objects as a group of pixels that are different from the group of pixels that form the background. The second subclass includes those targets that rely on optical flow [19–21].

When the camera is in a fixed position or the motion is very small and almost negligible, the background subtraction method works best, which is not the case for an onboard camera of a fast-moving aircraft. To overcome these problems, flow-based methods are used and give satisfactory results. But flow-based methods are critical when these come to detect the small and blurry objects. To overcome this problem, some methods use the combination of the both background subtraction and flow-based methods [22,23].

In our case, the background changes as the flying objects move through 3D spaces. The same types of flying objects (e.g., birds) can appear in various shapes and sizes simultaneously depending on the variable distance between the camera and the objects. So, it is clearly seen that only motion information is not sufficient to detect flying objects with high reliability. As said earlier, there are other methods which combine both background subtraction and flow-based methods, for example, it is seen that the work done in the references [24–27] depend on optical flow critically, which is again combined with [21] and may suffer from the low quality of flow vectors. Additionally, it can be included that the assumption made by Narayana et al. [26] about the translational camera motion is again violated in aerial videos.

4.2.3 Hybrid methods

These approaches combine information related to both object appearance and motion patterns and are therefore highly reliable. This method performs well regardless of the frame background variation and motion cue. Walk et al.'s [28] work can be shown as an example here, where histograms of flow vectors are used are used as features combined with the appearance features, which are more standard, and then fed to a statistical learning method. This approach is taken in Ref. [29], where at first the patches are aligned to compensate for motion and after that the differences of the frames (both consecutive and separate frames) are used as additional features. This alignment depends on the Lucas–Kanade optical flow algorithm [20]. The resulting algorithm responds very well when it comes to pedestrian detection and performs better than the single-frame methods.

The flow estimates are not reliable while detecting the smaller target objects which are harder to detect, and, like purely flow-based methods, this method becomes less effective.

The first deep learning object detection technique was Overleaf Network [30]. This technique used CNNs for classifying every part of the image along with a sliding window to detect and classify an image as object (object of interest) or nonobject (background). These earlier methods led to even more advanced techniques for object detection. In recent years, the deep learning community has proposed several object detectors, such as Faster RCNN [31], YOLO [32], R-FCN [33], SSD [34], and RetinaNet [35]. The primary focus of these algorithms was:

- to improve the accuracy and confidence score of the algorithm in terms of map; and
- to improve the computational complexity, so that these algorithms can be used effective for real-time object detection for embedded and mobile platforms [36].

These detection algorithms can be divided into two subclasses based on their high-level architecture: (1) single based approach and (2) two-step approach, that is, the region-based approach. The single-step approach is given more priority when the aim is to achieve faster results and high memory efficiency. Whereas the two-step approach is taken when accuracy is more important than time complexity and memory consumption.

4.2.4 Single-step detectors

Single-step detectors are simpler and faster than region-based models with reasonable accuracy. These detectors directly predict the coordinate offsets and object classes instead of predicting objects/nonobjects. SSD, YOLO, and RetinaNet are some of the examples of single-step detectors.

4.2.5 Two-step detectors/region-based detectors

This method has two stages for object detection:

- In the first stage, a set of regions are generated depending on the probability of an object being present within that particular region.
- In the second step, that particular area/region with high probability of having an object is fed as the input for the detection and classification of objects.

R-CNN, Fast R-CNN, FPN, Faster R-CNN, and R-FCN are some of the examples of single-step detectors.

4.3 Methodology

4.3.1 Model training

Here, a pretrained model of SSD [37] "Caffe Model" [38] is used for training the dataset. The dataset contains classes of flying objects like birds and aero planes (all types of planes are labeled as aero plane). OpenCV 4 library is used here. Within this OpenCV library, a highly improved deep learning model "DNN" is present. This "DNN" module supports a deep learning framework like "Caffe." The "Caffe" model architecture, which is used in this chapter, is shown below (Fig. 4.2):

The layers of Caffe model are described as follows:

- Vision layers:
  - Particularly operate to some region of the input and as a result, produce a corresponding region of the output.
  - Other layers ignore the spatial structure of the input. Though there are few exceptions.
  - May include convolution, pooling, Local Response Normalization layers.
  - Convolutional layers consist of a set of learnable filters which are applied over image.
- Loss layers:
  - Compares an output to a target and assigns cost to minimize.
  - May include SoftMax, sum-of-square/Euclidean, Hinge/Margin, information gain, accuracy score evaluation.

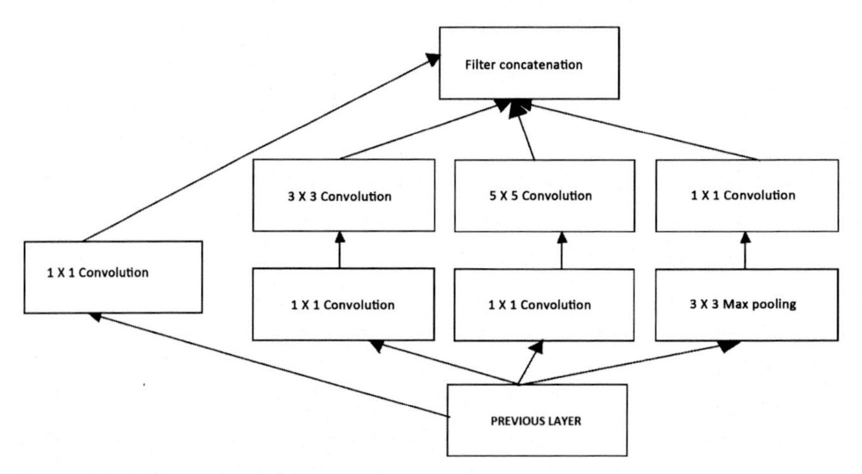

Figure 4.2 Caffe model architecture.

- Activation layers:
  - Elementwise operation is performed, taking one bottom blob and producing one top blob of the same size.
  - Transforms the weighted sum of inputs that goes into artificial neurons.
  - These functions are nonlinear to encode to encode complex patterns.
  - May include ReLU or Leaky ReLU, Sigmoid, TanH/Hyperbolic Tangent, Absolute Value, etc.
- Data layers:
  - Takes data directly from memory, or from files on disk (when efficiency is not critical) or common image formats.
  - May include database, In-memory, HDF5 input, HHDF5 output, images, windows or dummy.
- Common layers:
  - May include inner product, splitting, flattening, reshape, concatenation, slicing, elementwise operation, argmax, softmax, mean-variation normalization.

4.3.2 Evaluation metric

A confidence score is calculated as an evaluation standard. This confidence score shows the probability of the image being detected correctly by the algorithm and is given as a percentage. The scores are taken on the mean average precision at different IoU (Intersection over Union) thresholds. For this work, the thresholds are set at 0.6, 0.65, 0.7, 0.75, 0.8, 0.85, 0.9, that is, if the confidence of the detected object is over 60%, 65%, 70%, 75%, 80%, 85%, and 90%, respectively, then only the label will be taken for evaluation.

Intersection over Union (IoU) [39] is measured by the magnitude of overlap between two bounding boxes of two objects. The formula for IoU is given below:

$$\mathrm{IoU} = \frac{\text{Area of Overlap}}{\text{Area of Union}} \tag{4.1}$$

Eq. (4.1) can be expresses as:

$$\frac{\mathrm{IoU} = X \cap Y}{X \cup Y} \tag{4.2}$$

Precision is calculated at the threshold value depending on the True Positives (TP), False Positives (FP), and False Negatives (FN) which are the result of comparing the predicted object to the ground truth object. FN is not applicable for our case. A TP is counted when the ground truth bounding box is matched with the prediction bounding box. A FP is counted when a predicted object had no association with the ground truth object. A FN is counted when a ground truth object had no association with the predicted object. The confidence score is given as the mean over all the precision scores for all thresholds. The average precision is given as:

$$\text{Confidence} = \text{Average precision} = \frac{1}{|\text{Thresholds}|}\sum_{T}\frac{\text{TP}}{\text{TP}+\text{FP}+\text{FN}} \tag{4.3}$$

4.4 Results and discussions

Images of the flying objects are taken with a diverse background to test if the algorithm can work properly. Figs. 4.3 and 4.4 show the output when these images are fed to the algorithm for testing purposes. The output contains bounding boxes with a label of the object and a confidence score. This open source tool is known as Bounding Box Label [12], and is used to label all the different types of birds and aeroplanes (Table 4.1).

The average confidence score was calculated as the mean of all confidence scores given (for labeled images) by the algorithm. However, for testing purposes 100 images were provided for each class. It is seen that there are some images where the algorithm was detecting birds as aeroplanes and aeroplanes as birds. For some images instead of having aeroplanes and birds in the image, the network was not able to detect all the objects. The confusion matrix for each class is given below:

Tables 4.2 and 4.3 show that 97% aeroplane images and 95% bird images were correctly detected by the algorithm. For Table 4.2, TP = 97, FP = 1, FN = 2, True Negative (TN) = Not Applicable (NA). For Table 4.3, TP = 96, FP = 3, FN = 1, TN = NA.

4.4.1 For real-time flying objects from video

Figs. 4.5 and 4.6 are the screenshots taken from two videos. The first video shows the birds starting to fly and the second video is from an aeroplane takeoff.

Figure 4.3 Flying aeroplane detected by SSD algorithm. *SSD*, Single shot detection.

Figure 4.3 (Continued)

At first, the confidence of the network was tested for different objects in multiobject scenarios. Then the same network was applied to evaluate the performance of the network from a video feed. Some snapshots were taken after testing the video with the algorithms. The results from Figs. 4.5 and 4.6 clearly show that the network is working very well even for instantaneous object detection. The SSD algorithm was able to detect

Figure 4.4 Flying birds detected by SSD algorithm. *SSD*, Single shot detection.

and classify an object accurately in the video feed, even when the full object contours were obscured partially by another object. Based on the accuracy attained in detecting and classifying flying objects, this can be applied both for commercial and military applications. With slide

Figure 4.4 (Continued)

Table 4.1 Average confidence score for detected objects.

Object	Average confidence score
Bird	96.65
Aeroplane	97.89

Table 4.2 Confusion matrix for aero plane.

Classification	Class	Detected	
		Aeroplane	Not aeroplane
Actual	Aeroplane	97	2
	Not aeroplane	1	NA

NA, Not applicable.

Table 4.3 Confusion matrix for bird.

Classification	**Class**	**Detected**	
		Bird	**Not bird**
Actual	Bird	96	1
	Not bird	3	NA

NA, Not applicable.

Figure 4.5 Frames from flying birds video.

Figure 4.6 Frames from aeroplane takeoff video.

modifications, many transportation-related projects can apply this approach successfully.

4.5 Conclusion

Vision-based unmanned autonomous vehicles have a wide range of operations in intelligent navigation systems. In this chapter, a framework for real-time flying object detection is presented. The flying objects were detected by the network with a very high accuracy (average 97%, for bird and aeroplane classes) as shown in the results. These results show that the network is highly reliable for detecting flying objects. Here “SSD,” which is an open source platform for object detection and classification, is based on CNN. Future work can focus on detecting more flying objects like drone, helicopters, and even skydivers and paragliders. UAVs and CNN can be used for peripheral flying objects counting as well as detection and classification. Thus SSD can be very useful for detecting real-time objects. We hope that this work will add some value to modern ongoing research work on UAVs. Though this chapter focuses mainly on flying object detection for UAVs, the modified algorithm can be applied in the transportation and civil engineering fields. It is highly recommended to test the approach with a combination of different object classes on various images and to observe how these advancements will have an impact on how UAVs implement complex task. Another application can include automated identification of roadway features. These applications have the ability to transform the infrastructure of transportation asset management in the near future.

References

[1] Mercedes-Benz Intelligent Drive, http://techcenter.mercedes-benz.com/en/intelligentdrive/detail.html/.
[2] Mobileeye Inc. http://us.mobileye.com/technology/.
[3] G. Conte, P. Doherty, An integrated UAV navigation system based on aerial image matching, in: Proceedings of the IEEE Aerospace Conference (2008) 3142–3151.
[4] C. Martínez, I.F. Mondragon, M. Olivares-Méndez, P. Campoy, On-board and ground visual pose estimation techniques for UAV control, J. Intell. Robot. Syst. 61 (1–4) (2011) 301–320.
[5] L. Meier, P. Tanskanen, F. Fraundorfer, M. Pollefeys, PIXHAWK: a system for autonomous flight using onboard computer vision, in: Proceedings of the IEEE International Conference on Robotics and Automation (2011).

[6] C. Hane, C. Zach, J. Lim, A. Ranganathan, M. Pollefeys, Stereo depth map fusion for robot navigation, in: Proceedings of the International Conference on Intelligent Robots and Systems (2011) 1618–1625.
[7] S. Weiss, M. Achtelik, S. Lynen, M. Achtelik, L. Kneip, M. Chli, et al., Monocular vision for long-term micro aerial vehicle state estimation: a compendium, J. Field Robot. 30 (2013) 803–831.
[8] S. Lynen, M. Achtelik, S. Weiss, M. Chli, R. Siegwart, A robust and modular multi-sensor fusion approach applied to mav navigation, in: Proceedings of the Conference on Intelligent Robots and Systems, (2013).
[9] C. Forster, M. Pizzoli, D. Scaramuzza, SVO: fast semi-direct monocular visual odometry, in: Proceedings of the International Conference on Robotics and Automation (2014).
[10] T. Zsedrovits, A. Zarandy, B. Vanek, T. Peni, J. Bokor, T. Roska, Visual detection and implementation aspects of a uav see and avoid system, in: Proceedings of the European Conference on Circuit Theory and Design (2011).
[11] P. Felzenszwalb, R. Girshick, D. McAllester, D. Ramanan, Object Detection with Discriminatively Trained Part Based Models, IEEE Trans. Pattern Anal. Mach. Intell. (2010).
[12] F. Bastien, P. Lamblin, R. Pascanu, J. Bergstra, I. Goodfellow, A. Bergeron, et al., Theano: new features and speed improvements, 2012.
[13] A. Bosch, A. Zisserman, X. Munoz, Image classification using random forests and ferns, in: Proceedings of the International Conference on Computer Vision (2007).
[14] P. Dollar, Z. Tu, P. Perona, S. Belongie, Integral channel features, in: Proceedings of the British Machine Vision Conference (2009).
[15] N. Oliver, B. Rosario, A. Pentland, A Bayesian computer vision system for modeling human interactions, IEEE Trans. Pattern Anal. Mach. Intell. 22 (8) (2000) 831–843.
[16] N. Seungjong, J. Moongu, A new framework for background subtraction using multiple cues, in: Proceedings of the Asian Conference on Computer Vision. Springer Berlin Heidelberg (2013) 493–506.
[17] A. Sobral, BGSLibrary: an OpenCV C++ background subtraction library, in: Proceedings of the IX Workshop de VisaoComputacional, (2013).
[18] D. Zamalieva, A. Yilmaz, Background subtraction for the moving camera: a geometric approach, Comput. Vis. Image Underst. 127 (2014) 73–85.
[19] T. Brox, J. Malik, Object segmentation by long term analysis of point trajectories, in: Proceedings of the European Conference on Computer Vision (2010) 282–295.
[20] B. Lucas, T. Kanade, An iterative image registration technique with an application to stereo vision, in: Proceedings of the International Joint Conference on Artificial Intelligence (1981) 674–679.
[21] T. Brox, J. Malik, Large displacement optical flow: descriptor matching in variational motion estimation, IEEE Trans. Pattern Anal. Mach. Intell. (2011).
[22] Y. Zhang, S.-J. Kiselewich, W.-A. Bauson, R. Hammoud, Robust moving object detection at distance in the visible spectrum and beyond using a moving camera, in: Proceedings of the Conference on Computer Vision and Pattern Recognition (2006).
[23] S.-W. Kim, K. Yun, K.-M. Yi, S.-J. Kim, J.-Y. Choi, Detection of moving objects with a moving camera using non-panoramic background model, Mach. Vis. Appl. 24 (2013) 1015–1028.
[24] S. Kwak, T. Lim, W. Nam, B. Han, J. Han, Generalized background subtraction based on hybrid inference by belief propagation and bayesian filtering, in: Proceedings of the International Conference on Computer Vision (2011) 2174–2181.
[25] A. Elqursh, A. Elgammal, Online moving camera background subtraction, in: Proceedings of the European Conference on Computer Vision (2012) 228–241.

[26] M. Narayana, A. Hanson, E. Learned-miller, Coherent motion segmentation in moving camera videos using optical flow orientations, in: Proceedings of the International Conference on Computer Vision (2013).
[27] A. Papazoglou, V. Ferrari, Fast object segmentation in unconstrained video, in: Proceedings of the International Conference on Computer Vision (2013).
[28] S. Walk, N. Majer, K. Schindler, B. Schiele, New features and insights for pedestrian detection, in: Proceedings of the Conference on Computer Vision and Pattern Recognition (2010).
[29] D. Park, C.L. Zitnick, D. Ramanan, P. Dollar, Exploring weak stabilization for motion feature extraction, in: Proceedings of the Conference on Computer Vision and Pattern Recognition (2013).
[30] A.Z. Ryan, An overview of emerging results in cooperative UAV control, in: Proceedings of the 43rd IEEE Conference on Decision and Control (Vol. 1) (2004) 602–607.
[31] S.K.H. Ren, Faster R-CNN: towards real-time object detection with region proposal networks, IEEE Trans. Pattern Anal. Mach. Intell. 39 (6) (2017) 1137–1149.
[32] J. Redmon, S. Divvala, You only look once: unified, real-time object detection, in: Proceedings of the IEEE Conference on Computer Vision and Pattern Recognition (CVPR) (2016) 779–788.
[33] J. Dai, Y. Li, R-FCN: object detection via regionbased fully convolutional networks. NIPS, 2016, pp. 379–387.
[34] W. Liu, D. Anguelov, D. Erhan, C. Szegedy, S. Reed, C.-Y. Fu et al. SSD: single shot multibox detector. ECCV, 2016, pp. 21–37.
[35] T.-Y. Lin, P. Goyal. Focal loss for dense object detection. ICCV, 2017.
[36] S. Mandal, S. Biswas;V.E. Balas, R.N. Shaw, A. Ghosh, Motion prediction for autonomous vehicles from lyft dataset using deep learning, in: Proceedings of the 2020 IEEE 5th International Conference on Computing Communication and Automation (ICCCA) 30–31 Oct. (2020) 768–773, doi: 10.1109/ICCCA49541.2020.9250790.
[37] S. Mandal, V.E. Balas, R.N. Shaw, A. Ghosh, Prediction analysis of idiopathic pulmonary fibrosis progression from OSIC dataset, in: Proceedings of the 2020 IEEE International Conference on Computing, Power and Communication Technologies (GUCON), 2–4 Oct. (2020) 861–865, doi: 10.1109/GUCON48875.2020.9231239.
[38] M. Kumar, V.M. Shenbagaraman, A. Ghosh, Predictive data analysis for energy management of a smart factory leading to sustainability. Book Chapter [ISBN 978-981-15-4691-4] in: M.N. Favorskaya, S. Mekhilef, R.K. Pandey, N. Singh (Eds.), Innovations in Electrical and Electronic Engineering, Springer, 2020, pp. 765–773.
[39] S. Tripathi, B. Kang, Low-complexity object detection with deep convolutional neural network for embedded systems, Proc. SPIE 10396 (2017). 10 396 10 396 15.

CHAPTER FIVE

Depression detection for elderly people using AI robotic systems leveraging the Nelder–Mead Method

Anand Singh Rajawat[1], Romil Rawat[1], Kanishk Barhanpurkar[2], Rabindra Nath Shaw[3] and Ankush Ghosh[4]
[1]Department of Computer Science Engineering, Shri Vaishnav Vidyapeeth Vishwavidyalaya, Indore, India
[2]Department of Computer Science and Engineering, Sambhram Institute of Technology, Bengaluru, India
[3]School of Electrical, Electronics & Communication Engineering, Galgotias University, Greater Noida, India
[4]School of Engineering and Applied Sciences, The Neotia University, Kolkata, India

5.1 Introduction

IoT is a connected system of devices, combined with electronics, sensors allowing the collection and exchange of data between these objects [1–3]. Robotics is one of the most rapidly adopted domains in healthcare industries [4,5]. This phenomenon has been demonstrated by the significant enhancement of the efficiency and effectiveness of a healthcare system by incorporating robotics functions into a medical unit [6,7]. The societies' age continuously growing due to advancement in technology and this particular advancement is responsible for depression among all the age groups. Therefore, the provision of instant healthcare services has been perceived as a precondition for identifying user health situations [8]. The involvement of robotics in the detection of diseases and periodic monitoring of patients increased the life expectancy of patients. The most traditional experimental methods such as epigenetics, bionics, and neuropsychology research have been enhanced by new developments in robotics [9].

In the modern civilization, one-third of the total population is suffering from depression. Depression affects day to day activities and the quality of life of patients. Thus, a range of robotics-enabled applications is a viable solution for better healthcare for patients with such conditions. The

Artificial Intelligence for Future Generation Robotics.
DOI: https://doi.org/10.1016/B978-0-323-85498-6.00006-X

robotics-enabled technologies used in patients' body or in our living environments collect data from networked sensors leading to a successful change within the health landscape [10]. The aim of this article is to provide an outline of the subject and to suggest designing new robotics technologies related to the care of elderly disabled people in the area of depression and mental health. Similar studies are carried out focused on the work on robotics issues in the health care system [11] and on work into obstacles and limitations to wearable patient management systems embraced by acute and general clinicians [12]. We are also highlighting the impact of robotics (social animals and humanoids) to minimize depression in elderly people.

This chapter will be structured as follows. The second section of the chapter examines several methods for elderly people depression detection using robotics. In the third section, we have outlined related work and Section 5.4 consists of the elderly people depression signs and symptoms. Additionally, Section 5.5 highlights the proposed methodology and experimental results with some critical findings that will be debated. Finally, the conclusions and proposals for further research are provided.

5.2 Background

The global life expectancy index has been improved in recent decades, and is expected to rise in the future due to advances in medical research and associated technology. The number of senior adults (age >65) is increasing at an accelerated rate (see Fig. 5.1). Senior adults require further care and diagnosis, because irreparable harm can be done by small injuries or by trivial

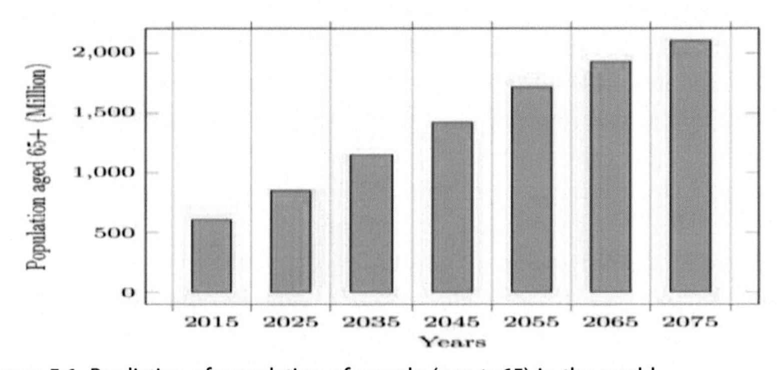

Figure 5.1 Prediction of population of people (age >65) in the world.

diseases [28]. Many older adults can also live alone while caregivers or medical experts or elderly people must monitor or be assisted with depression treatment. There is also an increasing need for advanced technology to provide the reliable care facilities for remote elderly depression. Hence, various contemporary approaches will be used to address the needs of the elderly despite their disabilities in daily life. One of the key advantages of robotics technology is that it allows continuous (i.e., 24/7) remote surveillance services to support the quality of life. Several attempts are made to use the robotics-based surveillance and treatment program, most of them focused from a specific perspective on some areas of elderly demand (for example, postpartum depression, depressive disorder, seasonal love, psychotic depressions, etc.). In view of the value of remote elder surveillance and the number of possible resources these systems can provide, a user-centric analysis is still missing. A user-focused architecture uses multiple channels to implement a framework that relies on the users' abilities and desires. Therefore it is critical for the current management frameworks for senior people to be taken into consideration from a user-focused point of view and for user-friendly design challenges to track applications. In this research, we study and classify the existing literature from an elderly person point of view to develop a comprehensive view of the field and to evaluate future trends. From a particular point of view, we address current problems and possible future robotics programs. Our research paves the way for studies to boost the quality of life of seniors. Inclusion of the latest robotics technologies in health has many benefits with respect to tracking, healthcare measures, and alerting and intelligence systems.

5.3 Related work

De la Torre Díez et al. [3] proposed an approach for monitoring the health of the people. For illnesses such as mental well-being, the quality and efficacy of medical is important for improving patient life [4]. The system was given for 30 days in a living room where close to 15 persons aged 68–78 years were tracked continually to support the success of unusual case detecting and forecasts for the seriousness of safety. For further classifying patients from stable people [5] the method suggested uses the combinatorial Fuzzy K-Nearest Neighbor and case-based Logic Classifier. The new healthcare network is designed to offer an urgent warning to the patient when an abnormality is identified. The method is

Table 5.1 Comparative study table different studies to detect depression using different domains.

Study reference	Research gap	Review topic
Chen et al. [14]	Not involvement of grey literature	Robotics
Pino et al. [15]	Smaller number of participants	Robotics
Kim et al. [16]	Smaller number of participants	Machine learning
Abdollahi et al. [17]	Smaller number of participants	Robotics
Zhongzhi et al. [18]	Not enough data on patient's hospitalization.	Deep learning
Xezonaki et al. [19]	Not enough data on patient's hospitalization	Fusion deep learning
Kargar and Mahoor [20]	Method for precise mood evaluation	Robotics
Bennett et al. [21]	Smaller number of participants	Robotics

Table 5.2 Comparison of different feature extracted for depression detection.

Study	Feature extraction	Machine learning technique/deep learning technique	Accuracy
Yalamanchili et al. [22]	Speech processing	Support Vector Machine (SVM)	93%
Zhu et al. [23]	EEG and eye movements	K-Nearest Neighbour (KNN)	78.37%
De Melo et al. [24]	Facial expressions	Convolutional Neural Network (3D)	80.6%
Zheng et al. [25]	Text embeddings	Dilated Temporal CNN and Causal Temporal CNN	95.3%
Guo et al. [26]	Facial expressions	Convolutional Neural Network (CNN)	78.83%
Huang et al. [27]	Speech-based features	Natural Processing Language techniques	82.9%

tested on the dataset of UCI-Parkinson and the findings indicate that the system works better than standard approaches [6]. So the living environment can be dynamically adjusted to assess the mood through the Happimeter. In our experiments, the quality of life of the elderly people can be improved by detecting disorders. Sau and Bhakta [13] aimed at developing a suitable predictive model for the diagnosis, using machine learning technology, of anxiety and depression among older patients, from social demographic and health factors. A data collection of 510 geriatric patients was used to assess and check 10 classifiers using 10 cross-validation approaches (Tables 5.1–5.4).

Table 5.3 Comparison of different types of robot forms were used to diminish depression symptoms.

S. no.	Study reference	Robot name	Root form of robot	Face scale results
1.	Bennett et al. [21]	Paro	Seal (animal)	4.17
2.	Kargar and Mahoor [20]	eBear	Bear (animal)	3.89
3.	Abdollahi et al. [17]	Ryan	Humanoid	3.67
4.	Pino et al. [15]	NAO	Humanoid	4.33

Table 5.4 Comparison of different classification algorithm [Random Forest Classifier, KNN, SVM, Deep belief networks (DBN), and MFCNN].

Model	Preprocessing Time (ms)	Fitting Time (ms)	Predicting Time(ms)	Accuracy
RFC	10.201384544372559	29.746789	11.72477769	65.80
KNN	10.114953526154300	29.622349	11.63286946	84.9
SVM	10.52247345546559	29.334456	7.853654576	87.71
DBNs	9.922475355263440	27.666456	8.635866485	73.40
MFCNN	9.312273455353449	28.634669	10.633666587	91.38

5.4 Elderly people detect depression signs and symptoms

The understanding of the signs and symptoms starts with an elderly person. Red flags include depression, sadness, or misery, mysterious or exacerbated pains and sorrows, lack of confidence in socializing, loss of weight or appetite, feelings of exhaustion or inability, determination or energy deficit, sleep issues (sleep challenges, sleep disorders, over-sleeping, or sleepiness in the morning), loss of self-worth (individual care), sluggish movement or voice, alcohol use or other prescription drugs, emphasis on mortality, depression, difficulty of memory, negligence for personal treatment (skipping appointments, medical recklessness/carelessness), lack of self-respect.

5.4.1 Causes of depression in older adults

We experience major changes in life that can increase a chance of depression as we grow older. This can include concerns with hygiene. All may lead to illness, such as disease and sickness, chronic or extreme discomfort, neurological impairment, and harm to the physical image by surgery and sickness. Depression may lead to such causes as living alone, a deteriorating social process induced by death or travel, diminished mobility due to disease, or the loss of driving privileges. The sense of intent has been that. Renaissance may lead to identity loss, reputation, self-confidence and financial stability, and the risk of depression is increased. You can also have physical limitations on things you used to enjoy. They include fear of illness or mortality and concern over financial or health issues.

5.4.2 Medical conditions that can cause elderly depression

Psychological response to the disease in older adults and elderly people: any chronic medical conditions can cause depression or make your depression symptoms worse, especially if it is painful, disabling, or life-threatening (Parkinson's disease).

5.4.3 Elderly depression as side effect of medication

Depression effects can also occur with other widely used medications as a side effect. If you take several drugs, you are particularly at risk. While prescription pharmaceutical products can impact anyone in the mood, the metabolization and the absorption of medications becomes less successful in our bodies. Speak to the doctor if you feel upset at the launch of a new medication.

5.4.4 Self-help for elderly depression

The human brains are never stopping evolving. Depression can be resolved if you try new ideas, learn to respond to change, keep physically and socially engaged, and interact with your friends and those you love. Even minor changes will affect how you feel greatly. By taking incremental steps day after day, the signs of depression will be better and more active, and optimistic aging will take place. Lifestyle changes in depression include a system that increases physical activity, suggests finding a new hobby or interest, suggests visiting family or friends regularly, emphasizes getting enough sleep every day, reading, and eating a well-balanced diet.

5.5 Proposed methodology

More advanced wearables, particularly those like Apple Watch, can also coordinate with critical implanted tech like continuous glucose monitors to make sure patients with diabetes are within safe blood sugar parameters. A few mental health startups are testing this model, essentially by allowing users to examine their mood, thought patterns, and then, based on this information, guiding them through cognitive behavioral therapy skills. Certain modes of treatment lend themselves better to the AI-based approach than others. Natural language processing is not yet at a point that it can handle complex conversation. However, many apps provide valuable skills-based strategies for managing anxiety and depression. If this can help people who would otherwise avoid all mental healthcare better manage their moods, then it is a step in the right direction. Digital mental healthcare is still an emerging field, but there are already a number of powerful tools helping to reach patients who otherwise may not have access to care. Mental health may be harder to measure in an objective sense, but that doesn't mean we can't use similar tools to improve our current model of care.

5.5.1 Proposed algorithm

Algorithm: Announcement or Alert generation

Input: Robotics sensor data (collected robotics device data with the help of body area network)

Output: Health status

Step 1: Compute the total score that defines the value in the Robotics device parameter (Check Total Unified Parkinson Disease Rating Scale (UPDRS) score)

Step 2: If (Checked the Score of is = 0) Check the net value.

Step 3: Check the status of the health = if people is Healthy;

Step 4: Monitor the Robotics Data Predict the results in terms of Time;

Step 5: (Computer the total score defines the value in the Robotics device parameter (Check Total _ Depression Rating Scale)}

Step 6: Else If (Check Total _ Depression Rating Scale > 1 and < 20), do {

Step 7: Check the status of the health = Compute the value of Mild Parkinson;

Step 8: If that condition is generated then produce the alert automatically (Alarm)

Step 9: provide the required precaution for that

Step 10: Monitor the Robotics Data Predict the results in term of Time;

Step 11: Again compute the (Computer the total score define the value in the Robotics device parameter (Check Total _ Depression Rating Scale);

}

Step 12: Else If (Check Check Total _ Depression Rating Scale > 21 and < 35), do {

Step 13: Check the status of the health = if people is Healthy;

Step 14: Alarm produce the emergency notification

Step 15: monitoring the all records (older Adult)

Step 16: Display features after T time period;

Step 17: Again compute the (Computer the total score define the value in the

Robotics device parameter (CheckCheck Total _ Depression Rating Scale)

}

Step 18: Check the status of the health = if people are Healthy;

Step 19: Alarm produce the emergency notification;}

Elderly-centered robotics and automation-based remote control: given the expected high pace of elderly population growth (see Fig. 5.1) in the immediate future, substantial efforts are required to leverage emerging concepts and innovations such as robotics and automation in the elderly care market. A number of approaches have been given to meet the concerns of the elderly by compensating for the deficit or reducing the eventual impacts. Patients were analyzed based upon their text information, audio, video, and image (robotics-based data) recording. Each of these techniques is being used to extract the features of the elderly people and the features are analyzed by using various classification models, including a robotics depression model that can be used to determine the rate of depression in a patient. Current technologies and facilities are seen in Fig. 5.2. For this aim, several small, medium, and big initiatives were often launched to meet the concerns of older adults for various objectives. The program includes remote control capabilities as well as resources including updates (daily, weekly, etc.), recommendations, and early alarms. It also communicates the knowledge of the aged to third party personnel (i.e., nurses, medical professionals, and ambulance units) so that they may participate in an incident, offer treatment recommendations, and provide assistance. In fact, the device is able to obtain input from third party representatives to give the customer more flexibility and enhance efficiency (e.g., responsiveness and accuracy of the method). Various problems should be addressed in these schemes, such as data validity, data reliability, and data security [29,31]. Below we research

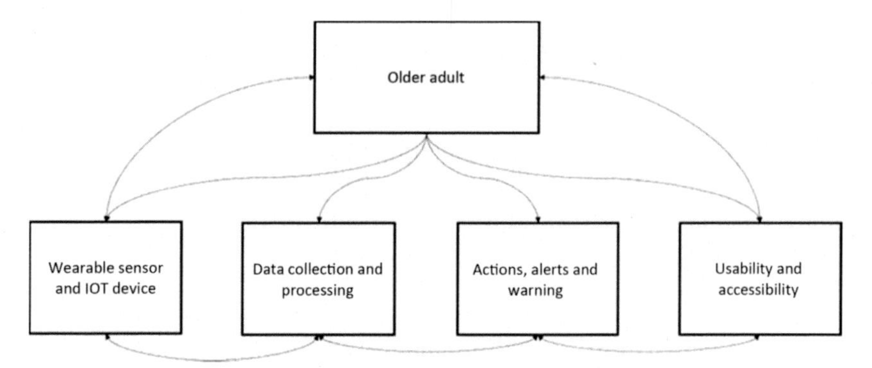

Figure 5.2 Working of proposed methodology.

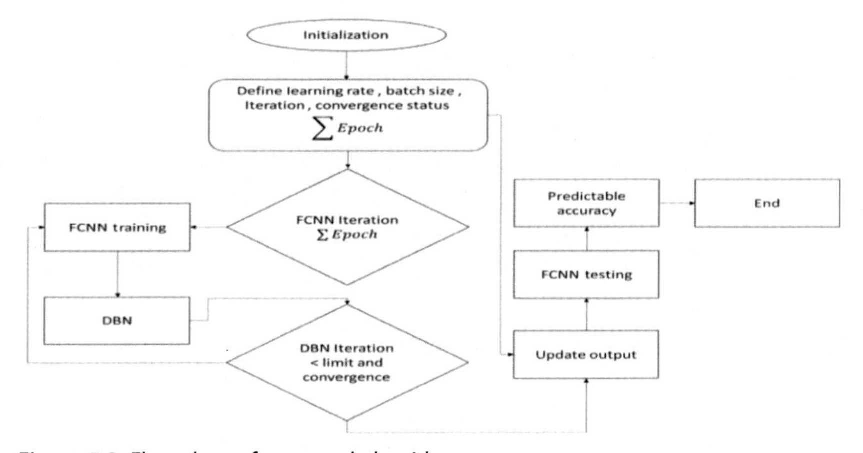

Figure 5.3 Flow chart of proposed algorithm.

systems and solutions relevant to robotics and automation by classifying them into five groups. It should be noted that some of those solutions are in more than one category and we have depicted the flow chart of proposed algorithm, fusion model and data collection methods (Figs. 5.3–5.5).

5.5.2 Persistent monitoring for depression detection

Impairments and recurrent conditions can only be treated adequately by continuous supervision. For example, diabetes forms should be tracked during life. Thus the life span, invasiveness, and battery life of the sensor systems should be known for long-term operation. Most physical and emotional disorders can occur by long-term observation of the user's behaviors and behavior. Collecting and processing incoming large data

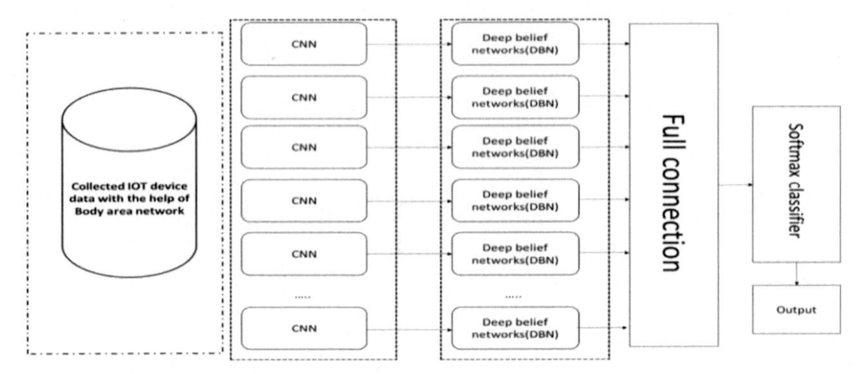

Figure 5.4 Fusion CNN (CNN + DBN).

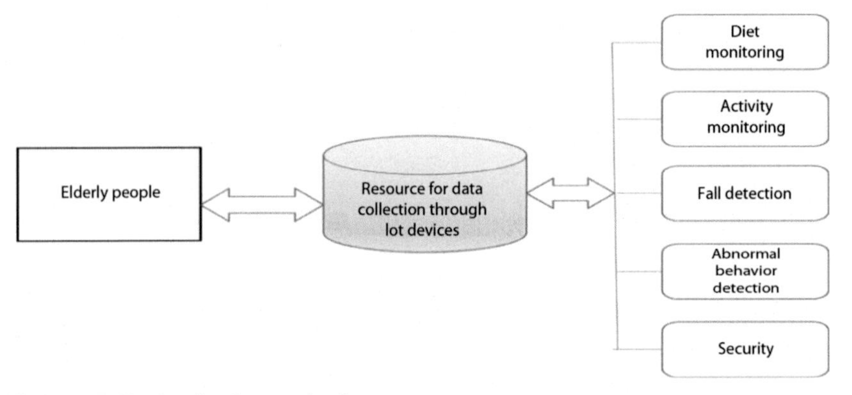

Figure 5.5 Data collection methods.

from a user's everyday life over a long period of time, including physical exercise, feeding, sleeping, and social life, will help the program derive useful information for the analysis of depression data. This enduring awareness can often include improvements in behaviors and could be required for medical examination. Specific psychiatric disorders are observed by long-term analysis of behavioral adjustments.

5.5.3 Emergency monitoring

In older people, the risk of medical incidents (e.g., stroke and heart attack) and environmental injuries (e.g., slip detection, loss) is higher. A robust video surveillance of elderly people also needs to identify emergency incidents and respond appropriately (i.e., alert nurses or medical professionals to reduce the consequences). In fact, it increases device stability in the event of Internet access not being accessible.

5.5.4 Personalized monitoring

Personalized monitoring is one of the essential objectives to be considered in a comprehensive monitoring process by robotics and automation-based older people monitoring systems. In general, many presumptions for the device are defined depending on common user specifications and conditions. Such presumptions contribute to inefficiencies in long-term supervision of the elderly. Hence, adaptability and customization techniques (i.e., personalization) are of utmost importance. The principle of self-awareness [31] and the big data analytics can be incorporated into the robotics and automation-based frameworks to include a customized control mechanism.

5.5.5 Feature extraction

There are numerous ways to undertake abstract feature extraction, which is the most important aspect to develop a effective model. Also, the features are inferior illustrations of audio signals, which display the speech characteristics by depressed elderly people living in old age homes and hospitals. The audio samples collected from datasets are divided into periodic time interval of around 2–3 seconds each. A spectrogram is a graphic representation of audio signals, demonstrating the amplitude of the audio of a signal over a specific time interval. In Fig. 5.6, the y-axis shows frequency and the y-axis contains time (in seconds). The red patches

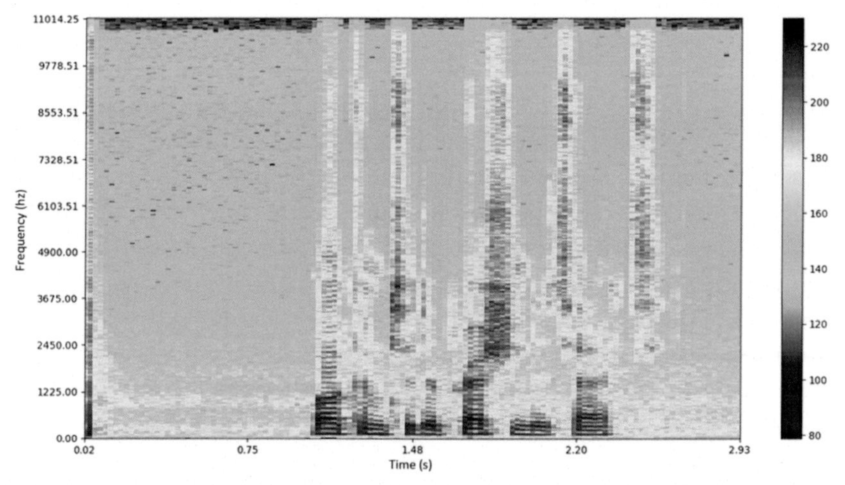

Figure 5.6 Acoustic feature extraction from audio sample collected from depression data-set.

observed in the spectrogram depict the high difference changes in the amplitude.

5.6 Result analysis

Big data analytics, on the one side, derive valuable knowledge from incoming heterogeneous big data to render the system aware of the patient and underlying conditions of the world, and, on the other side, self-conscious methods allow the system to optimize its actions in relation to the condition (i.e., patient state and environmental state) and to change attention to crucial parameters in the system. For example, the program identifies multiple goals for various medical conditions dependent on the elderly illnesses in the sense of safety surveillance. The goals specify the degree of priority of the parameters. In other terms, for each parameter they determine the data collection speeds from the sensors, the period of execution, and the order of the data review. Identifying distress in patients using fog computing is a modern method of enhancing the level of healthier living for elderly adults with minimal treatment expenses and services.

Elderly people accused of developing depression can determine whether or not they're suffering from depression by voice recordings. The people can record their voice samples and upload them to the cloud over the internet after registering with the web or mobile application. By analyzing their speech samples, voice samples, text content, audio, video, and images (robotics-based data), the classifier categorizes the individual into either depression or safe. If the individual is identified as Parkinson's then the patient's cell phone should be notified instantly. Professionals or clinicians may remotely analyze the cloud speech samples, speech notes, text content, audio, video, and images (robotics-based data) and administer the prescription. The online platform holds all customer records and acts as a repository for both healthcare consumers and providers. Our proposed smart network is planned to take advantage of both cloud and fog computing, such as local data processing, alert generation, sorting, data protection, accessibility, and central storage. Furthermore, influential features of their speech samples, speech notes, text content, audio, video, and picture (robotics-based data) characterize patients with depression by using MFCNN combinatorial method. The statistical parameters like

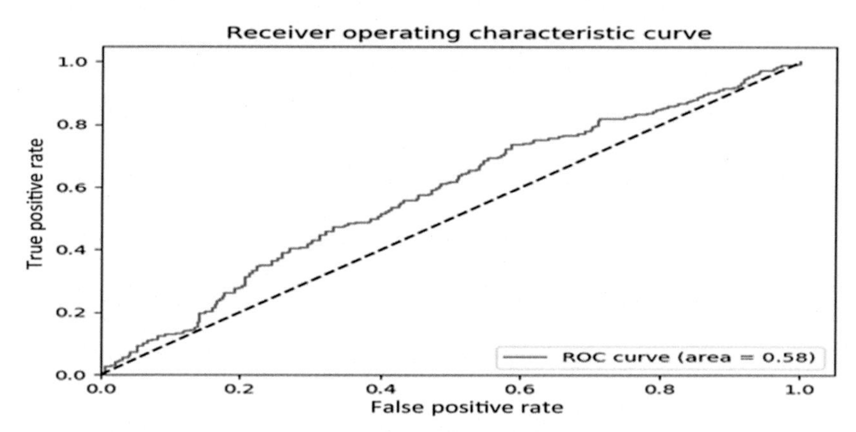

Figure 5.7 Receiver Operating Curve for MFCNN model.

weighted mean and spearman correlation are used to measure the efficiency of the machine learning models like Decision Tree, KNN, SVM, Deep belief networks (DBN), and MFCNN. The classification accuracy produced on the Parkinson dataset by the proposed MFCNN classifier is 91.7.

The Receiver Operating Curve (ROC) is widely used to plot the binary classifier system. In Fig. 5.7, the ROC is used for TP and FP (where TP stands for True Positive and FP stands for False Positive) for the MFCNN model and the total area calculated for the curve is 0.58. True positive is an outcome of the correct prediction by the MFCNN model and False Positive is the test result which is wrongly predicted by the MFCNN model.

IT and science integration, such as medical informatics, could change patient treatment as we know it, lowering prices and saving lives. The main focus of this analysis was to offer a summary of advanced work on robotics and automation programs, technologies and frameworks in the identification of elderly people utilizing robotics to enhance robotics-based healthcare systems. The research also indicates the advantages of robotics and automation in the diagnosis of depression as well as the technologies and frameworks built to increase the quality of life of the individual with this form of condition. There are several forms of depression, including postpartum depression, depressive depression, seasonal affective disorder, and psychotic depression, to name a few. But, if undetected they will take a toll on life to a certain level and individuals suffering from depression will make poor choices to end their lives. The up-to-date infrastructure contributes to deficits in the country's economic development. Efficient strategies for identifying and curing mental illnesses also need to be developed and can touch society so that the shame surrounding psychiatric illness diminishes and society may solve their everyday difficulties and be successful. Any tools and instruments may be used to identify depression, dependent on their study. It is feasible to use sensors and tools and some of the latest secure robotics and automation to build a platform that will help people diagnose distress and therefore see physicians and therapists. In the future, we will implement chatbots that may also be used for depression diagnosis and rehabilitation. Various combined technology may contribute to an efficient depression diagnosis approach. We have proposed a model for finding depression in elderly people which provides the highest accuracy and optimal predicting time and preprocessing time. In the further steps, we are developing a humanoid based-robotic model which is capable of detecting and preventing symptoms of depression in elderly people.

References

[1] M.M. Baig, S. Afifi, H. Gholam Hosseini, F. Mirza, A systematic review of wearable sensors and IoT-based monitoring applications for older adults – a focus on ageing population and independent living, J. Med. Syst. 43 (8) (2019). Available from: https://doi.org/10.1007/s10916-019-1365-7.

[2] I. Azimi, A.M. Rahmani, P. Liljeberg, H. Tenhunen, Internet of things for remote elderly monitoring: a study from user-centered perspective, J. Ambient Intell. Humaniz. Comput. 8 (2) (2016) 273–289. Available from: https://doi.org/10.1007/s12652-016-0387-y.

[3] I. De la Torre Díez, S.G. Alonso, S. Hamrioui, E.M. Cruz, L.M. Nozaleda, M.A. Franco, IoT-based services and applications for mental health in the literature, J. Med. Syst. 43 (1) (2018). Available from: https://doi.org/10.1007/s10916-018-1130-3.

[4] A. Manocha, R. Singh, M. Bhatia, Cognitive intelligence assisted fog-cloud architecture for generalized anxiety disorder (GAD) prediction, J. Med. Syst. 44 (1) (2019). Available from: https://doi.org/10.1007/s10916-019-1495-y.

[5] M. Devarajan, L. Ravi, Intelligent cyber-physical system for an efficient detection of Parkinson disease using fog computing, Multimed. Tools Appl. (2018). Available from: https://doi.org/10.1007/s11042-018-6898-0.

[6] A. Capodieci, P. Budner, J. Eirich, P. Gloor, L. Mainetti, Dynamically adapting the environment for elderly people through smartwatch-based mood detection, 2018. Available from: https://doi.org/10.1007/978-3-319-74295-3_6.

[7] S. Dalal, S. Jain, M. Dave, A systematic review of smart mental healthcare (December 29, 2019), Proceedings of the 5th International Conference on Cyber Security & Privacy in Communication Networks (ICCS) 2019 | National Institute of Technology, Kurukshetra, India. Available at SSRN: <https://ssrn.com/abstract = 3511013> or <https://doi.org/10.2139/ssrn.3511013>.

[8] M. Helbich, Y. Yao, Y. Liu, J. Zhang, P. Liu, R. Wang, Using deep learning to examine street view green and blue spaces and their associations with geriatric depression in Beijing, China, Environ. Int. 126 (2019) 107–117. Available from: https://doi.org/10.1016/j.envint.2019.02.013.

[9] E. Lin, P.H. Kuo, Y.L. Liu, Y.W. Yu, A.C. Yang, S.J. Tsai, A deep learning approach for predicting antidepressant response in major depression using clinical and genetic biomarkers, Front. Psychiatry 9 (2018) 290. Available from: https://doi.org/10.3389/fpsyt.2018.00290.
[10] B. Farrow, and S. Jayarathna, Technological advancements in post-traumatic stress disorder detection: a survey. 2019 IEEE 20th International Conference on Information Reuse and Integration for Data Science (IRI), 2019. Available from: https://doi.org/10.1109/iri.2019.00044.
[11] R. Smith, S. Meeks, Screening older adults for depression: barriers across clinical discipline training, Innov. Aging 3 (2) (2019). Available from: https://doi.org/10.1093/geroni/igz011.
[12] M.V. Costa, M.F. Diniz, K.K. Nascimento, K.S. Pereira, N.S. Dias, L.F. Malloy-Diniz, et al., Accuracy of three depression screening scales to diagnose major depressive episodes in older adults without neurocognitive disorders, Rev. Bras. Psiquiatr. 38 (2) (2016) 154–156. Available from: https://doi.org/10.1590/1516-4446-2015-1818.
[13] A. Sau, I. Bhakta, Predicting anxiety and depression in elderly patients using machine learning technology, Healthc. Technol. Lett. 4 (6) (2017) 238–243. Available from: https://doi.org/10.1049/htl.2016.0096.
[14] S.C. Chen, C. Jones, W. Moyle, Social robots for depression in older adults: a systematic review, J. Nurs. Scholarsh. (2018) 50. Available from: https://doi.org/10.1111/jnu.12423.
[15] O. Pino, G. Palestra, R. Trevino, B. Carolis, The humanoid robot NAO as trainer in a memory program for elderly people with mild cognitive impairment, Int. J. Soc. Robot. (2019). Available from: https://doi.org/10.1007/s12369-019-00533-y.
[16] J. Kim, N. Liu, H. Tan, C. Chu, Unobtrusive monitoring to detect depression for elderly with chronic illnesses, IEEE Sens. J. 17 (17) (2017) 5694–5704. Available from: https://doi.org/10.1109/JSEN.2017.2729594.
[17] H. Abdollahi, A. Mollahosseini, J.T. Lane, and M.H. Mahoor, A pilot study on using an intelligent life-like robot as a companion for elderly individuals with dementia and depression, 2017 IEEE-RAS 17th International Conference on Humanoid Robotics (Humanoids), 2017. Available from: https://doi.org/10.1109/humanoids.2017.8246925.
[18] X. Zhongzhi, Q. Zhang, W. Li, M. Li, P.S.F. Yip, Individualized prediction of depressive disorder in the elderly: a multitask deep learning approach, Int. J. Med. Informatics (2019) 103973. Available from: https://doi.org/10.1016/j.ijmedinf.2019.103973.
[19] D. Xezonaki, G. Paraskevopoulos, P. Alexandros, S. Narayanan, Affective conditioning on hierarchical networks applied to depression detection from transcribed clinical interviews, 2020.
[20] B.A.H. Kargar and M.H. Mahoor, A pilot study on the eBear socially assistive robot: Implication for interacting with elderly people with moderate depression, 2017 IEEE-RAS 17th International Conference on Humanoid Robotics (Humanoids), 2017. Available from: https://doi.org/10.1109/humanoids.2017.8246957.
[21] C.C. Bennett, S. Sabanovic, J.A. Piatt, S. Nagata, L. Eldridge, and N. Randall, A robot a day keeps the blues away, 2017 IEEE International Conference on Healthcare Informatics (ICHI), Park City, UT, 2017, pp. 536–540, Available from: https://doi.org/10.1109/ICHI.2017.43.
[22] B. Yalamanchili, , N.S. Kota, M.S. Abbaraju, V.S.S. Nadella, and S.V. Alluri, Real-time acoustic based depression detection using machine learning techniques, 2020 International Conference on Emerging Trends in Information Technology and Engineering (ic-ETITE), 2020. Available from: https://doi.org/10.1109/ic-etite47903.2020.394.

[23] J. Zhu, Z. Wang, T. Gong, S. Zeng, X. Li, B. Hu, et al., An improved classification model for depression detection using EEG and EyeTracking Data, IEEE Trans. Nanobiosci. (2020) 1. Available from: https://doi.org/10.1109/tnb.2020.2990690.

[24] W.C. De Melo, E. Granger, and A.cHadid, Combining global and local convolutional 3D networks for detecting depression from facial expressions, 2019 14th IEEE International Conference on Automatic Face & Gesture Recognition (FG 2019), 2019. Available from: https://doi.org/10.1109/fg.2019.8756568.

[25] W. Zheng, L. Yan, C. Gou, and F.-Y. Wang, Graph attention model embedded with multi-modal knowledge for depression detection, 2020 IEEE International Conference on Multimedia and Expo (ICME), 2020. Available from: https://doi.org/10.1109/icme46284.2020.9102872.

[26] W. Guo, H. Yang, and Z. Liu, Deep neural networks for depression recognition based on facial expressions caused by stimulus tasks, 2019 8th International Conference on Affective Computing and Intelligent Interaction Workshops and Demos (ACIIW), 2019. Available from: https://doi.org/10.1109/aciiw.2019.8925293.

[27] Z. Huang, J. Epps, D. Joachim, V. Sethu, Natural language processing methods for acoustic and landmark event-based features in speech-based depression detection, IEEE J. Sel. Top. Signal Process. (2019) 1. Available from: https://doi.org/10.1109/jstsp.2019.2949419.

[28] WHO Library Cataloguing-in-Publication Data International statistical classification of diseases and related health problems. - 10th revision, Fifth edition, 2016.

[29] M. Jara, V. Adaui, B.M. Valencia, D. Martinez, M. Alba, C. Castrillon, et al., Real-time PCR assay for detection and quantification of Leishmania (Viannia) organisms in skin and mucosal lesions: exploratory study of parasite load and clinical parameters, J. Clin. Microbiol. Jun. 51 (6) (2013) 1826–1833. Available from: https://doi.org/10.1128/JCM.00208-13.

[30] R. Agarwal, V. Dhar, Editorial - Big Data, Data Science, and Analytics: The Opportunity and Challenge for IS Research, Inf. Syst. Res. 25 (2014) 443–448.

[31] B. Henze, A. Dietrich, An approach to combine balancing with hierarchical whole-body control for legged humanoid robots, IEEE Robotics and Automation Letters (RA-L) 1 (2) (2016) 700–707.

CHAPTER SIX

Data heterogeneity mitigation in healthcare robotic systems leveraging the Nelder–Mead method

Pritam Khan, Priyesh Ranjan and Sudhir Kumar
Department of Electrical Engineering, Indian Institute of Technology Patna, India

6.1 Introduction

Robotic systems are gaining importance with the progress in artificial intelligence. The robots are designed to aid mankind with accurate results and enhanced productivity. In the medical domain time-series data and image data are used for research and analysis purposes. However, heterogeneity in data from various devices processing the same physical phenomenon will result in inaccurate analysis. Robots are not humans, but are trained by humans, therefore the implementation of the data heterogeneity mitigation technique in the robotic systems will help in improving the performance. In IoT (Internet of Things) networks, heterogeneity is a challenge and various methods are designed for mitigating the problem. The machine learning algorithms that are used for classification, regression, or clustering purposes, are based on the data that are the input to the model. Therefore the data from different devices can yield different classification accuracies.

6.1.1 Related work

Data heterogeneity and its mitigation have been explored in few works. Jirkovzky et al. [1] discuss the various types of data heterogeneity present in a cyberphysical system. The different categories of data heterogeneity are syntactic heterogeneity, terminological heterogeneity, semantic heterogeneity, and semiotic heterogeneity [1]. In that work, the causes of the heterogeneity are also explored. The device heterogeneity is considered

Artificial Intelligence for Future Generation Robotics.
DOI: https://doi.org/10.1016/B978-0-323-85498-6.00012-5

for smart localization using residual neural networks in [2]. However, heterogeneity mitigation in the data used for localization would have generated more consistency in the results. Device heterogeneity is addressed using a localization method and Gaussian mixture model by Pandey et al. [3]. Zero-mean and unity-mean features of Wi-Fi (wireless fidelity) received signal strength used for localization assist to mitigate device heterogeneity in Refs. [4] and [5]. The approach is however not used for data classification or prediction purposes from multiple devices using neural networks. The work presented in Ref. [6] aims at bringing interoperability in one common layer by using semantics to store heterogeneous data streams generated by different cyberphysical systems. A common data model using linked data technologies is used for the purpose. The concept of service oriented architecture is introduced in Ref. [7] for mitigation of data heterogeneity. However, the versatility of the data management system remains unexplored.

In this work, we leverage the Nelder–Mead (NM) method for heterogeneity mitigation in the raw data from multiple sources. Mitigation of data heterogeneity from multiple sources will increase the reliability of robotic systems owing to the consistency in prediction, classification, or clustering results. We classify normal and abnormal electrocardiogram (ECG) signals from two different datasets. Although the ECG signals in the two datasets belong to different persons, for the sake of proving the benefit of heterogeneity mitigation, we classify the normal and abnormal ECG signals using Long Short Term Memory (LSTM) from two different datasets.

6.1.2 Contributions

The major contributions of our work are as follows:

1. Data heterogeneity is mitigated, thereby increasing the consistency in performance among various devices processing the same physical phenomenon.
2. The NM method is a proven optimization technique which validates our data heterogeneity mitigation process.
3. A robotic system classifies the normal and abnormal ECG signals with consistency in classification accuracies after mitigating data heterogeneity.

The rest of the paper is organized as follows. In Section 6.2, we discuss the preprocessing of data from two datasets and mitigate the heterogeneity using the NM method. The classification of the ECG data is discussed in Section 6.3. The results are illustrated in Section 6.4. Finally, Section 6.5 concludes the paper indicating the scope for future work.

6.2 Data heterogeneity mitigation

We leverage the NM method for mitigating the data heterogeneity to help the robotic systems yield consistent results from different data sources. This in turn helps the users to increase the reliability of the modern robotic systems. Prior to mitigation of heterogeneity, we need to preprocess the data which can be from different sources.

6.2.1 Data preprocessing

We consider time-series healthcare data analysis in a robotic system. The raw data from various sources can contain unequal numbers of samples owing to different sampling rates and different devices. Therefore we equalize the number of sample-points for each time-series data from different sources. This can be implemented by using the upsampling or downsampling technique. Upsampling is preferred over downsampling to avoid any information loss. The number of sample points from different devices corresponding to a particular time-series data is upsampled to that of the device generating the maximum number of sample points. Once the data from each device contains an equal number of samples, we apply the NM method of heterogeneity mitigation.

6.2.2 Nelder–Mead method for mitigating data heterogeneity

The NM method used for mitigating the instant based intersample distance is robust to noisy signals [8,9]. Using this technique eliminates the necessity of denoising the signals separately in the processing stage. Additionally, this method does not require the knowledge of gradients. The method can be applied when the functions are not locally smooth. NM method minimizes a function of n variables by comparing the function values at $(n + 1)$ vertices of a general simplex and then replacing the vertex with the highest value by another point [9]. An explanation of the NM method with figures is provided in Ref. [9]. A simplex is a generalized figure which can be a point, a line, a triangle, a tetrahedron or any higher dimensional convex hull. If we have initial points $x_{k,\ 1}$, $x_{k,\ 2}$,..., $x_{k,\ n}$ from n devices initially, then after NM minimization we have transformed points $x'_{k,\ 1}$, $x'_{k,\ 2}$,..., $x'_{k,\ n}$ respectively.

Let us illustrate the NM method used with an example of three sample data points $x_{k,\ 1}$, $x_{k,\ 2}$, and $x_{k,\ 3}$ obtained from three devices. Let the corresponding functions at the three vertices be $f\ (x_{k,\ 1})$, $f\ (x_{k,\ 2})$, and $f\ (x_{k,\ 3})$. In the first iteration, one is assigned to be the worst vertex (say, $x_{k,\ 1}$), another to be second worst (say, $x_{k,\ 2}$), while the other is the best (say, $x_{k,\ 3}$). We can write this as: $f\ (x_{k,\ 1}) = \max\ if\ (x_{k,\ i})$, $f\ (x_{k,\ 2}) = \max_{i\neq 1} f\ (x_{k,\ i})$, and $f\ (x_{k,\ 3}) = \min_{i\neq 1} f\ (x_{k,\ i})$. Then we calculate the centroid x_c for the best side which is opposite to the worst vertex $x_{k,\ 1}$.

$$x_c = \frac{1}{n} \sum_{i \neq 1} x_{k,i} \tag{6.1}$$

Now we find the modified simplex from the existing one. As this is a 3-simplex, we have a convex hull of four vertices, that is, a tetrahedron. We replace the worst vertex $x_{k,\ 1}$ by a better point obtained with respect to the best side. Reflection, expansion, or contraction of the worst vertex yields the new better point $x_{k,\ 1}$. The test-points are selected on the line connecting the worst point $x_{k,\ 1}$ and the centroid x_c. If the method achieves a better point with a test-point then the simplex is reconstructed with the accepted test-point. However, if the method fails then the simplex is shrunk toward the best vertex $x_{k,\ 3}$. Fig. 6.1A describes the

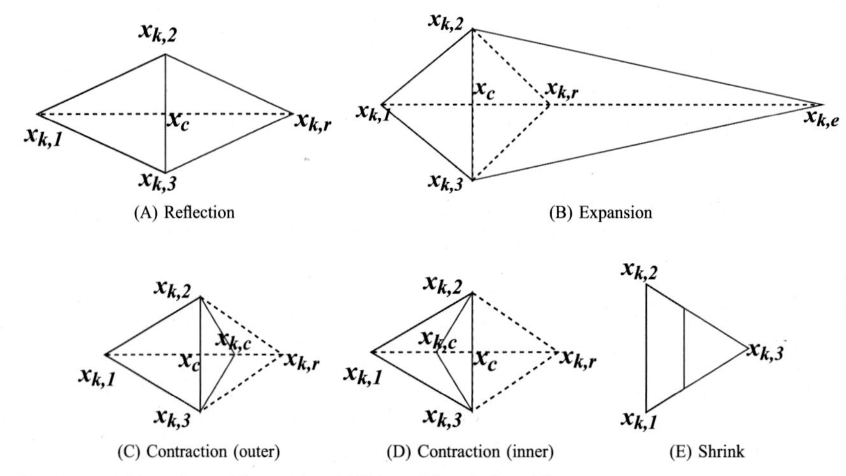

Figure 6.1 Transformation using Nelder–Mead algorithm.

reflection technique. The reflection point $x_{k,r}$ is calculated as:

$$x_{k,r} = x_c + \alpha(x_c - x_{k,1}), \text{ where } \alpha > 0 \quad (6.2)$$

The function corresponding to vertex $x_{k,r}$ is $f(x_{k,r})$. If $f(x_{k,3}) \leq f(x_{k,r}) < f(x_{k,2})$, then $x_{k,r}$ is accepted and the iteration is terminated. If $f(x_{k,r}) < f(x_{k,3})$, we go for expansion as shown in Fig. 6.1B. The expansion point is computed as:

$$x_{k,e} = x_c + \gamma(x_{k,r} - x_c), \text{ where } \gamma > 1 \text{ and } \gamma > \alpha \quad (6.3)$$

Now if $f(x_{k,e}) < f(x_{k,r})$, then $x_{k,e}$ is selected else $x_{k,r}$ is accepted and then iteration is terminated. Hence, a greedy minimization technique is followed. If $f(x_{k,r}) \geq f(x_{k,2})$, then the contraction point $x_{k,c}$ is found using the better point out of $x_{k,1}$ and $x_{k,r}$. If $f(x_{k,2}) \leq f(x_{k,r}) < f(x_{k,1})$, then contraction is outside the initially considered 3-simplex, as shown in Fig. 6.1C. The contracted point $x_{k,c}$ is given by:

$$x_{k,c} = x_c + \beta(x_{k,r} - x_c), \text{ where } 0 < \beta < 1 \quad (6.4)$$

If $f(x_{k,c}) \leq f(x_{k,r})$, then $x_{k,c}$ is accepted and iteration is terminated, else we go for the shrink method. However, if $f(x_{k,r}) \geq f(x_{k,1})$, then contraction is considered inside the initially considered 3-simplex as shown in Fig. 6.1D. In this case we get the contracted point $x_{k,c}$ as:

$$x_{k,c} = x_c + \beta(x_{k,1} - x_c), \text{ where } 0 < \beta < 1 \quad (6.5)$$

If $f(x_{k,c}) < f(x_{k,1})$, then $x_{k,c}$ is accepted and iteration is terminated, else we perform the shrink operation as shown in Fig. 6.1E. In the shrink method, we obtain all the new vertices as:

$$x_{k,i} = x_{k,3} + \delta(x_{k,i} - x_{k,3}) \text{ where } 0 < \delta < 1 \text{ and } i = 0, 1, 2 \quad (6.6)$$

Shrinking transformation is used when contraction fails due to the occurrence of a curved valley causing some point of the simplex to be much further away from the valley bottom than the other points. This transformation finally brings all the points into the valley. Based on the values of sample points $x_{k,1}$, $x_{k,2}$, and $x_{k,3}$ from three different devices, the NM method uses expansion, contraction, or shrinking to obtain a new set of transformed points $x'_{k,1}$, $x'_{k,2}$, and $x'_{k,3}$, respectively.

Here we illustrate the NM minimization using a 3-simplex, that is, data from three different devices. However, this can be generalized for any higher or lower order simplex. For example, if we consider two devices, then we have 2-simplex, that is, a triangle, and if we consider four devices, then we have a pentagon for evaluation.

The NM method must terminate after a certain period of time. Three tests are conducted to terminate the NM method [8,9]. If any of them is true then the NM method stops operating. The domain convergence test yields positive results if all or some vertices of the present simplex are very near to each other. The function-value convergence test is true if the functions at different vertices are very close to each other. Finally, the no-convergence test becomes positive if the number of iterations exceeds a certain maximum value.

6.3 LSTM-based classification of data

LSTM is a special category of recurrent neural network aimed at handling time-series data. We classify the ECG signals into normal and abnormal in this work using LSTM. Our objective is to show that there is consistency in classification accuracy for heterogeneity mitigated data over heterogeneous data in robotic systems. The computational complexity of LSTM is of the order of $O(1)$ per time step and weight. LSTM has a recurrently self-connected linear unit called "constant error carousel" (CEC). The recurrency of activation and error signals helps CEC provide short-term memory storage for the long term [10,11]. In an LSTM unit, the memory part comprises a cell and there are three "regulators" called gates, controlling the passage of information inside the LSTM unit: an input gate, an output gate and a forget gate [12]. LSTM works in four steps:

1. Information to be forgotten is identified from previous time step using forget gate.
2. New information is sought for updating cell state using input gate and tanh.
3. Cell state is updated using the above two gates information.
4. Relevant information is yielded using output gate and the squashing function.

In Fig. 6.2, LSTM-dense layer-softmax activation architecture is illustrated.

Let x_t be the input received by the LSTM cell in Fig. 6.2. i_t, o_t, and c_t represent input gate, output gate, and long-term memory of current

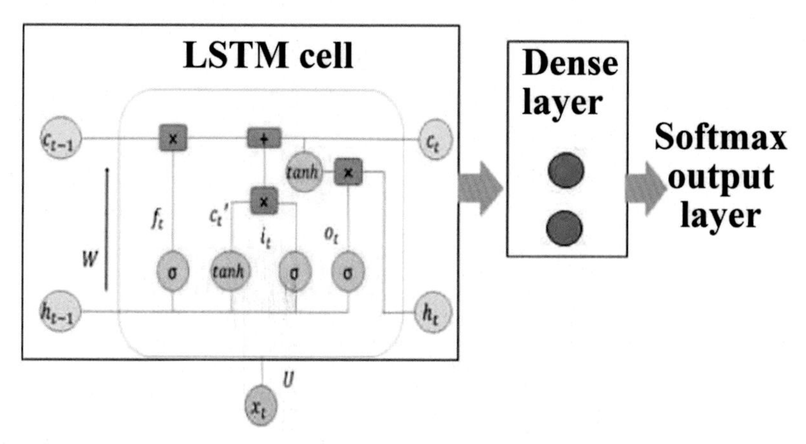

Figure 6.2 Softmax activated LSTM-dense layer architecture. *LSTM*, Long Short Term Memory.

time-step t, respectively. W and U are weight vectors for f, c, i, and o. The hidden unit (h_t) at each time-step learns which data to keep and which to discard. Accordingly, the forget gate (f_t) keeps a 0 (to forget entirely) or a 1 (to remember). Hence, the activation function in the forget gate is always a sigmoid function (σ).

$$f_t = \sigma\left(x_t \times U_f + h_{t-1} \times W_f\right) \tag{6.7}$$

Further, we require to evaluate the information learnt from xt where the activation function is generally tanh.

$$c_t = \tanh(x_t \times U_c + h_{t-1} \times W_c) \tag{6.8}$$

Now prior to addition of the information to memory, it is necessary to learn which parts of the information are not redundant and are required to be saved through the input gate i_t. Here also the activation function is σ.

$$i_t = \sigma(x_t \times U_i + h_{t-1} \times W_i) \tag{6.9}$$

Thus the working memory is updated and the output gate vector is learnt.

$$o_t = \sigma(x_t \times U_o + h_{t-1} \times W_o) \tag{6.10}$$

The present cell state is obtained as

$$c_t = f_t \times c_{t-1} + i_t \times c_t' \tag{6.11}$$

And the present hidden unit as

$$h_t = o_t \times \tanh(c_t) \tag{6.12}$$

In Fig. 6.2, LSTM cell output is given to a dense layer. The softmax activation function is given to the output stage after the dense layer. Softmax activation function for any input vector $\boldsymbol{y}$ comprising n elements is given by: $\text{softmax}(y_i) = \frac{e^{y_i}}{\sum_{j=1}^{n} e^{y_j}}$ where $\boldsymbol{y} = [y_1, \ldots, y_n]^T$.

In ECG data classification, information about the previous state is required in the present state for accurate classification of activities.

6.4 Experiments and results

We use the data from the Massachusetts Institute of Technology-Beth Israel Hospital (MIT-BIH) arrhythmia database and Physikalisch-Technische Bundesanstalt (PTB) Diagnostic ECG database to mitigate heterogeneity and further classify them into normal and abnormal categories [13–15]. An amalgamation of both the databases is carried out as it contributes to heterogeneity. We select those ECG waveforms from both the datasets with the R peaks at approximately the same positions by mapping. Finally, we train and test the LSTM with the selected parts of the MIT-BIH dataset and the PTB database. Next we mitigate the heterogeneity between MIT-BIH and PTBDB data using NM method as observed from Table 6.1. Heterogeneity mitigation leads to increase in accuracy during classification.

6.4.1 Data heterogeneity mitigation using Nelder–Mead method

One beat of an ECG waveform is chosen from both the datasets as shown in Fig. 6.3A, whose heterogeneity when removed gives the waveforms as shown in Fig. 6.3B corresponding to the NM method. We calculate the RMSE (root mean squared error) between the two ECG waveforms, as shown in Fig. 6.3A and B. Fig. 6.3B represents the RMSE for the same beat of ECG from two devices after heterogeneity mitigation using the NM method [8,9]. For Fig. 6.3A we have RMSE $= 2.390 \times 10^{-1}$. Using the NM method, we can reduce the RMSE to 1.008×10^{-5}. The RMSE is calculated for 8000 iterations and the time taken by the NM

Table 6.1 Performance comparison of different models on different data.

Different data with model evaluation parameters		LSTM + softmax
Homogeneous MIT-BIH arrhythmia data	Training accuracy%	98.11
	Test accuracy%	97.51
	Precision	0.9513
	Recall	0.9107
	F1 Score	0.9305
Homogeneous PTBDB ECG data	Training accuracy%	94.23
	Test accuracy%	93.30
	Precision	0.9356
	Recall	0.9343
	F1 score	0.9349
Heterogeneous data (MIT-BIH + PTBDB Data)	Training accuracy%	90.31
	Test accuracy%	86.18
	Precision	0.8850
	Recall	0.8850
	F1 score	0.8850
Heterogeneity mitigated data using Nelder–Mead method	Training accuracy%	91.84
	Test accuracy%	91.67
	Precision	0.9246
	Recall	0.9243
	F1 Score	0.9243

MIT-BIH, Massachusetts Institute of Technology-Beth Israel Hospital; *NA*, Not applicable.

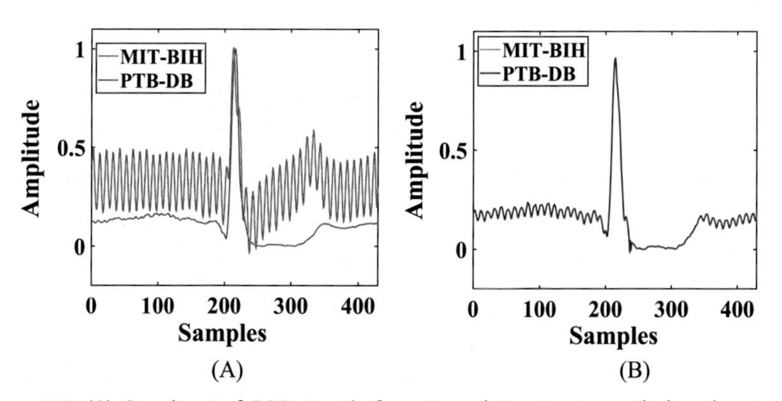

Figure 6.3 (A) One beat of ECG signals from two datasets upsampled and normalized. (B) Same beats of ECG from two datasets after heterogeneity mitigation using NM method. *ECG*, Electrocardiogram; *NM*, Nelder–Mead.

method for the same is 1.436 seconds. Notably, RMSE is very small since the signals are normalized between 0 and 1.

6.4.2 LSTM-based classification of data

Leveraging data from different sources lead to data heterogeneity. In this work, we use two different ECG datasets for classification where each dataset comprises data from different persons acquired using different devices of different specifications. We classify the ECG data from the two datasets namely MIT-BIH and PTB-DB separately as well as in combined states. We create a fusion of the normal classes from the two databases and similarly of the abnormal classes also, thereby creating a heterogeneous platform of data. The LSTM-softmax deep learning model is used for classification purpose. In order to achieve the best performance using the LSTM model, we tune the hyperparameters like number of LSTM hidden units and batch size, and regularize the model using the ℓ_2 norm of regularization. Both the number of hidden units and batch size are tuned to 64 for the LSTM model. The learning rate is selected as 0.01 for maximizing the accuracy of our model. Also, we select the rate for ℓ_2 norm of regularization as $\lambda = 0.001$.

The performance evaluation of the classification model based on test data is carried out using a confusion matrix [16]. We obtain the following performance metrics from the confusion matrix.

$$\text{Accuracy} = \frac{(\text{TP} + \text{TN})}{(\text{TP} + \text{TN} + \text{FP} + \text{FN})}$$

$$\text{Precision} = \frac{\text{TP}}{(\text{TP} + \text{FP})}$$

$$\text{Recall (Sensitivity)} = \frac{\text{TP}}{(\text{TP} + \text{FN})}$$

where TP, TN, FP, and FN represent True Positive, True Negative, False Positive, and False Negative, respectively. Another performance metric called F1 score is calculated from precision and recall values. F1 score is the weighted average of accuracy and recall and it varies from 0 to 1. It is defined as follows:

$$\text{F1 Score} = 2 \times \frac{(\text{Precison} \times \text{Recall})}{(\text{Precison} + \text{Recall})}$$

All the evaluation parameters, namely, precision, recall, and F1 score, are tabulated in Table 6.1 in addition to training and testing accuracies. From

Table 6.1, it is observed that the NM method-based heterogeneity mitigated data yields a training accuracy of 91.84% and a test accuracy of 91.67%, which are higher than that for heterogeneous data (MIT-BIH and PTB-DB combined). The heterogeneous data has a maximum training accuracy of 90.31% and a test accuracy of 86.18%. We also evaluate our model in terms of other performance metrics like precision, recall, and F1-score.

However, when we consider each dataset separately for classification of ECG signals into normal and abnormal separately, we get higher accuracy, as observed from Table 6.1.

Therefore when we implement the heterogeneity mitigation method, we trade-off accuracy to some extent at the expense of increasing reliability on the robotic system. A robotic system yielding the results as shown in Fig. 6.3B is more reliable and consistent compared to that in Fig. 6.3A.

6.5 Conclusion and future work

In this work, we mitigate the heterogeneity in the data that is analyzed by a robotic system. Mitigation of data heterogeneity is carried out using the NM optimization technique. It is observed from the results that the RMSE reduces by a large extent among the heterogeneity mitigated data as compared to heterogeneous data. Additionally, we classify the healthcare ECG data from two datasets, namely MIT-BIH and PTB-DB, into normal and abnormal categories. We use the LSTM-softmax deep learning model for classification purposes. The NM method-based heterogeneity mitigated data yield higher classification accuracies in comparison to the heterogeneous data.

We are carrying out further research to mitigate the heterogeneity for multivariate healthcare data in robotic systems. Most of the healthcare conditions require multiparameter monitoring. Therefore the robots need to be trained with algorithms capable of handling multivariate analysis with the data heterogeneity mitigated.

Acknowledgment

This work acknowledges the support rendered by the Early Career Research (ECR) award scheme project "Cyber-Physical Systems for M-Health" (ECR/2016/001532) (duration 2017–20), under Science and Engineering Research Board, Government of India.

References

[1] V. Jirkovský, M. Obitko, V. Mařík, Understanding data heterogeneity in the context of cyber-physical systems integration, IEEE Trans. Ind. Inform. 13 (2) (2016) 660–667.
[2] A. Pandey, P. Tiwary, S. Kumar, S.K. Das, Residual neural networks for heterogeneous smart device localization in iot networks, in: Proceedings of the 2020 29th International Con- ference on Computer Communications and Networks (ICCCN). IEEE, (2020) 1–9.
[3] A. Pandey, R. Vamsi, S. Kumar, Handling device heterogeneity and orientation using multistage regression for GMM based localization in IoT networks, IEEE Access. 7 (2019) 144. 354–144 365.
[4] S. Kumar, S.K. Das, ZU-mean: fingerprinting based device localization methods for IoT in the presence of additive and multiplicative noise, in: Proceedings of the Workshop Program of the 19th International Conference on Distributed Computing and Networking. ACM (2018) 15.
[5] S. Kumar, S.K. Das, Target detection and localization methods using compartmental model for internet of things, IEEE Trans. Mob. Comput. 19 (2019) 2234–2249.
[6] A. Kazmi, Z. Jan, A. Zappa, M. Serrano, Overcoming the heterogeneity in the internet of things for smart cities, in: Proceedings of the International Workshop on Interoperability and Open-Source Solutions. Springer (2016) 20–35.
[7] T. Fan, Y. Chen, A scheme of data management in the internet of things, in: Proceedings of the 2010 2nd IEEE International Conference on Network Infrastructure and Digital Content. IEEE (2010) 110–114.
[8] J.A. Nelder, R. Mead, A simplex method for function minimization, Comput. J. 7 (4) (1965) 308–313.
[9] S. Singer, J. Nelder, Nelder–Mead algorithm, Scholarpedia 4 (7) (2009) 2928.
[10] F.A. Gers, J. Schmidhuber, F. Cummins, Learning to forget: continual prediction with LSTM, 1999.
[11] F.A. Gers, N.N. Schraudolph, J. Schmidhuber, Learning precise timing with lstm recurrent networks, J. Mach. Learn. Res. 3 (2002) 115–143. no. Aug.
[12] S. Hochreiter, J. Schmidhuber, Long short-term memory, Neural Comput. 9 (8) (1997) 1735–1780.
[13] G.B. Moody, R.G. Mark, The impact of the MIT-BIH arrhythmia database, IEEE Eng. Med. Biol. Mag. 20 (3) (2001) 45–50.
[14] R. Bousseljot, D. Kreiseler, A. Schnabel, Nutzung der EKG-Signaldatenbank CARDIODAT der PTB über das internet, Biomed. Tech./Biomed. Eng. 40 (s1) (1995) 317–318.
[15] A.L. Goldberger, L.A. Amaral, L. Glass, J.M. Hausdorff, P.C. Ivanov, R.G. Mark, et al., Physiobank, physiotoolkit, and physionet: components of a new research resource for complex physiologic signals, Circulation 101 (23) (2000) e215–e220.
[16] S. Visa, B. Ramsay, A. Ralescu, E. Knaap, Confusion matrix-based feature selection. vol. 710, 01 (2011) 120–127.

CHAPTER SEVEN

Advance machine learning and artificial intelligence applications in service robot

Sanjoy Das[1], Indrani Das[2], Rabindra Nath Shaw[3] and Ankush Ghosh[4]

[1]Department of Computer Science, Indira Gandhi National Tribal University, Regional Campus Manipur, Imphal, Manipur
[2]Department of Computer Science, Assam University, Silchar, India
[3]Department of Electrical, Electronics & Communication Engineering, Galgotias University, Greater Noida, India
[4]School of Engineering and Applied Sciences, The Neotia University, Kolkata, India

7.1 Introduction

Artificial intelligence (AI) and machine learning (ML) applications are undergoing tremendous growth with endless uses in our real life [1]. Robotics engineering also attracts the attention of researchers from every corner of the world. Robotics is an interdisciplinary domain that includes Computer Science along with other applied mechanisms. Robotics involves the design, construction, and use of AI and ML techniques to perform certain specific tasks that are generally performed by human beings. The concepts of service robots assist human beings to perform various types of works with less error. The service robot may be semi or fully automatic in nature. A service robot assists in jobs that are dangerous, repetitive, dull, etc., in nature. For the betterment of human beings service robots are also included in manufacturing operations in automobile industries. Based on the situations they are capable of taking decisions to finish the task. The major scope of AI in robotics includes vision, grasping, motion control, data, etc. Vision means robots detect items which they have never seen before and recognize them. The robots grasp things they have never seen before, and with the help of AI and ML determine the best position to grasp a thing. ML techniques help robots with motion control so that they can interact in a dynamic environment and avoid any

Artificial Intelligence for Future Generation Robotics.
DOI: https://doi.org/10.1016/B978-0-323-85498-6.00002-2

obstacle in their way. The understanding of data patterns to robots is determined by AI and ML techniques and helps robots to act efficiently.

In this chapter, our major objectives are how service robots are helpful in home automation along with other application areas of real life. Why is the popularity of home service robots increasing day by day? Second, how are AI and ML techniques useful in efficiently managing the home service robot? We have shown various uses of ML and AI techniques in a robust way of managing in various fields. How businesses across the world are transforming their organization by including service robots is discussed. The advantages of service robot along with the applications of AI and ML are discussed.

Section 7.1 gives the basic history of robotics and its applications. Section 7.2 includes a literature review. Uses of AI and ML are discussed in Section 7.3. The chapter is concluded in Section 7.4 and the future scope is discussed in Section 7.5.

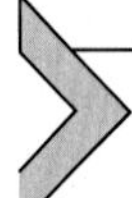

7.2 Literature reviews

7.2.1 Home service robot

The digital home concept has recently emerged. It is characterized by a network of sensors and actuators which provide services in terms of comfort, security, and energy management [2]. The robots generally reduce production costs and improve the precision and productivity in the tasks they are assigned to do. With the invention of robots, they substitute for the manpower of any organization in discharging many complicated tasks. The major motivation behind including robots is the reduction of overall production costs, minimizing production time, and producing flawless products [3].

Service robots have transformed homes, streets, schools, hospitals, manufacturing units, etc. The iRobot Corporation has taken some of the steps toward service robots penetrating into the home, via Roomba [3]. Roomba is an autonomous robotic vacuum cleaner that has sold more than 2 million units worldwide [3].

The major goal of robotics is to design intelligent machines. The purpose of the service robot is to assist human beings in discharging day-to-day works more effectively and efficiently.

Table 7.1 Classification of service robots and application area [3–5].

Types of robots	Application area
Professional service robots [3,5]	Hospitality, healthcare, warehouse, agriculture, etc. [3]
Personal service robots [3]	Servant robot [5], automated wheelchair [3]
Defense robots [3]	Demining robots [4,5], unmanned aerial vehicles [4,5] unmanned ground based vehicles [4,5]
Domestic robot [4,5]	Household works
Entertainment robot [4,5]	Toy robots multimedia/remote presence
Medical assistance [4,5]	Diagnostic systems [4], robot-assisted surgery [4,5]
Home security and surveillance [4]	Intruder detection [4]
Field robotics [4,5]	Agriculture field monitoring [4,5], milking robots [4,5], livestock monitoring [4]
Professional cleaning [4,5]	Floor, window, and wall cleaning [4,5]
Inspection and maintenance systems [4]	Tank, tubes, pipes, and sewers [5]
Construction and demolition [4,5]	Nuclear demolition [4] building construction [4,5]
Logistic systems [4,5]	Cargo handling [4]
Rescue and security applications [4,5]	Fire and disaster fighting robots [4,5]

The service robot is classified according to their works and the area where they are used. In Table 7.1, we have summarized the various kinds of service robots.

7.3 Uses of artificial intelligence and machine learning in robotics

In this section, we discuss how AI and ML couple with service robots [6] and their applications and advantages.

7.3.1 Artificial intelligence applications in robotics [6]

Nowadays most industries worldwide use AI techniques to enhance the performances of service robots. In this way they increase the overall efficiency of their systems. The industries like healthcare, construction, logistics, and management companies are using service robots to boost their various organizational day-to-day operations. Fig. 7.1 shows the various

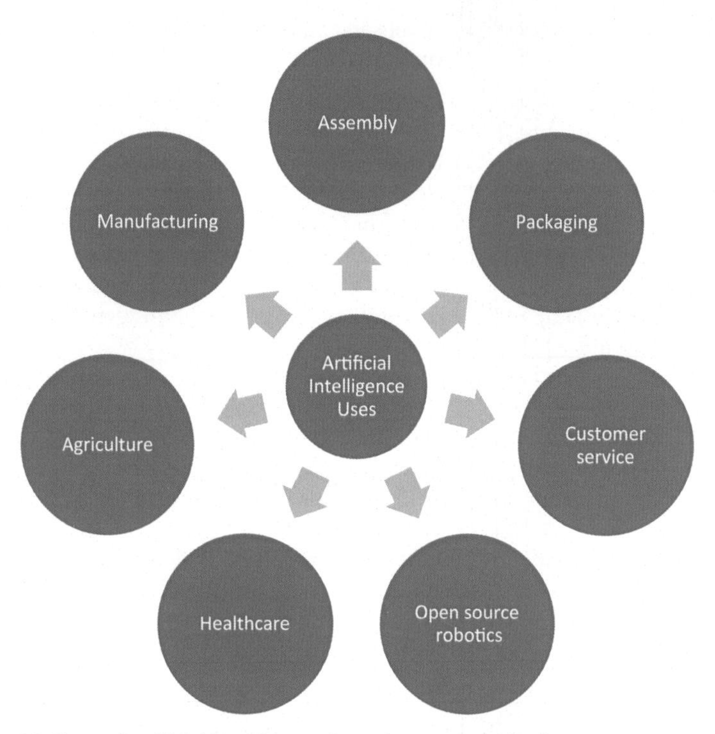

Figure 7.1 Uses of artificial intelligence in various sectors [6,7].

sectors where AI techniques are combined with services robots. With the help of AI techniques service robots are more adaptable to new environments and improve the overall performance of the system.

Assembly [7]

AI is a highly useful tool in robotic assembly applications. When combined with advanced vision systems, AI can help with real-time course correction, which is particularly useful in complex manufacturing sectors like aerospace. AI can also be used to help a robot learn on its own which paths are best for certain processes while it is in operation.

Packaging [7]

Robotic packaging uses forms of AI frequently for quicker, lower cost, and more accurate packaging. AI helps save certain motions a robotic system makes, while constantly refining them, which makes installing and moving robotic systems easy enough for anybody to do.

Customer service [7]

Robots are now being used in a customer service capacity in retail stores and hotels around the world. Like AI, a Chatbot is used to answer customer queries. They use AI natural language processing abilities to interact with customers. These machines are capable of learning from human beings and act accordingly.

Open source robotics [7]

A handful of robotic systems are now being sold as open source systems with AI capability. This way, users can teach their robots to do custom tasks based on their specific application, such as small-scale agriculture. The convergence of open source robotics and AI could be a huge trend in the future of AI robots.

Table 7.2 shows the various renowned organizations working in the robotics field and uses AI to make service robots more adaptable and scalable to various new situations [8].

The AI techniques have been proven to be very efficient and effective in robotics over the last few decades. Endless applications are surrounding us in real life, where AI is combined with robotics [9]. The industrial sectors use AI for efficient and accurate management of raw materials, polishing, cleaning, cutting, fixing, and precise drilling of various metallic and nonmetallic materials. The precise execution of activities increases the overall organization's error-free productivity and timely completion of various works.

Another premier sector is medical science, where the use of robotics is in high demand because of its precise and accurate execution of a task. Much typical surgery is nowadays performed with the help of robots. Many clinical examination works are carried through robots.

In the field of space, research robotics plays a vital role because where a human being cannot reach and survive robots can be used to collect data and exploration of the area.

7.3.2 Machine learning applications in robotics [10]

In Section 7.3.1 we have discussed the various uses of AI in robotics. In this section, we will discuss the uses and applications of ML techniques in robotics. The uses of ML have a tremendous impact in the field of robotics, because the precision, accuracy, and learning capabilities of robots are increasing day by day. In Fig. 7.2, various domains where ML applications are used in robotics are highlighted [6,10,11].

Table 7.2 Various project and their features with applications domain [8].

Name of project	Country	Features	Applications
Boston Dynamics	Massachusetts	Dynamic, intelligent, adaptive	Environment and terrains
Canvas Technology	Colorado	Autonomous, intelligent, easily adaptable	Unstructured warehouse processes for indoors and outdoors uses
Dronesense	Texas	Autonomous as well as manual	Public safety-fire service, emergency response, law enforcement, rescue operation
MISO Robotics	California	Autonomous	Commercial kitchens, flippy and thermal vision, decreases food waste
Neurala	Massachusetts	Autonomous and intelligent	Improve intelligence in cars, phones, drones and cameras etc.,
Rethink Robotics	Massachusetts	Collaborative and adaptive	Furniture company
Sea Machines	Massachusetts	Autonomous	Marine and maritime industry
Veo Robotics	Massachusetts	Autonomous, flexible	Car assembly lines
Perceptive Automata	Mass	Adaptive, autonomous	Autonomous vehicle with prediction of human behavior
Piaggio Fast Forward	Boston	Autonomous	Grocery shop for helping the customer
Uipath	New York	Autonomous and adaptive	Software company for perform repetitive tasks.
Engineered Arts	England	Humanoid and semihumanoid	Facial and object recognition
Cruise	San Francisco	Autonomous	Self-driving car
Hanson Robotics	China	Humanoid	Eye contact, facial recognition, speech and natural conversation.
Brain Corp	California	Adaptable and flexible	Navigate unstructured environments like large warehouses and store floors
Starship	San Francisco	Autonomous	Groceries, food delivery items.
Cloudminds	California	Controlled from the cloud	Airport patrol with vision and navigation capabilities
Vicarious	California	Adaptive very fast	Determine and verify human versus machine users, for example, CAPTCHA response

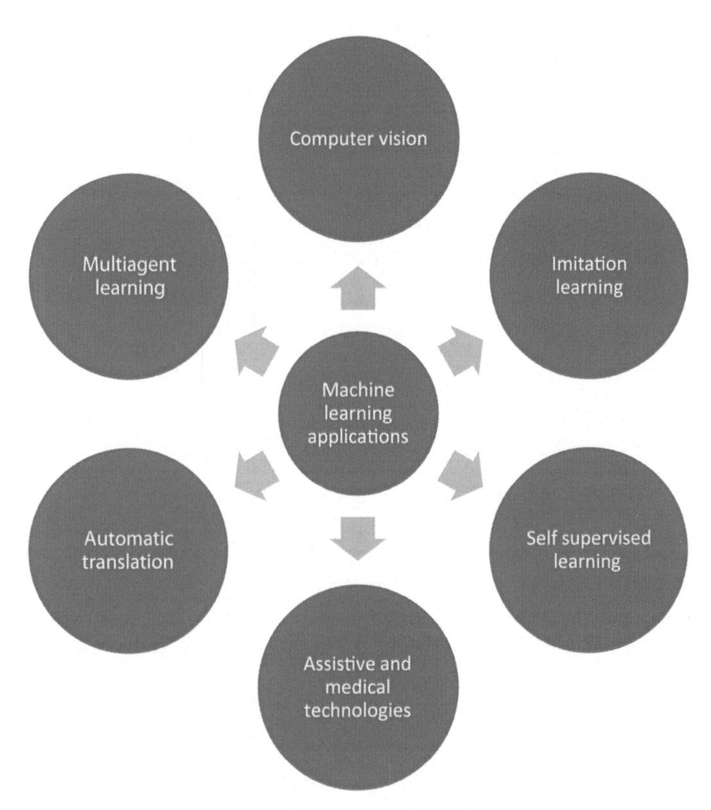

Figure 7.2 Machine learning uses in robotics [6].

Businesses worldwide are growing very rapidly and are transformative in nature. When ML is used in robotics, it further enhances the productivity of the system. There are endless domains where ML has proven to be very effective with robotics in managing things, for example, in fields like medical science, manufacturing, logistic management security and surveillance, quality control, etc. [6].

7.4 Conclusion

In this chapter, we have discussed many issues and uses of service robots in our real-life activities. With the help of AI, ML techniques can improve the efficiency of various types of service robots. Various fields of application from medical science, manufacturing, to logistic management are also discussed. The projects running based on AI and ML with robotics are summarized. AI and ML help in transforming conventional businesses across the globe with the help of robotics. The AI and ML techniques help robots to be more adaptable to a new situation and make them self learning and act accordingly. There are endless benefits to organizations when AI and ML are included with robotics. Many groundbreaking works continue and the possibilities are endless in the coming future. This chapter's purpose is to give a taste of the advantages of ML and AI in various applications that exist in robotics.

7.5 Future scope

The uses of robotics in our real life is endless and human beings are mostly benefiting. The tremendous growth in robotics is seen in many developing countries and it can ease human life by including service robots with AI and ML techniques. In the future, we will explore various other aspects, specifically the uses of robotics in medical sciences. The scope of robotics in the future is never-ending, this is just the beginning.

References

[1] W. Wang, K. Siau, Artificial intelligence, machine learning, automation, robotics, future of work and future of humanity: a review and research agenda, J. Database Manage. 30 (1) (2019) 61–79. Available from: https://doi.org/10.4018/JDM.2019010104.

[2] J. Ram, "Integration of Service Robots in the Smart Home," Universidad de Sevilla Departamento, 2016.

[3] J.R. De La Pinta, J.M. Maestre, E.F. Camacho, I.G. Alonso, Robots in the smart home: a project towards interoperability, Int. J. Ad Hoc Ubiquitous Comput. 7 (3) (2011) 192–201. Available from: https://doi.org/10.1504/IJAHUC.2011.040119.

[4] International Federation of Robotics, Classification of service robots by application areas, Introd. Serv. Robot. 2 (2012) 38.

[5] Service Robots," Robotics Industries Association, 2015. <https://www.robotics.org/service-robots/what-are-professional-service-robots#>:~:text = There are service robots that, automate dangerous or laborious tasks. (accessed: 30.12.20).

[6] 5 Applications of Machine Learning in Robotics. <https://www.usmsystems.com/applications-of-machine-learning-in-robotics/>. (accessed 30.12.20).

[7] M. Kumar, V.M. Shenbagaraman, A. Ghosh, Predictive data analysis for energy management of a smart factory leading to sustainability, in: Favorskaya, M.N., Mekhilef, S., Pandey, R.K., Singh, N. (Eds.) Innovations in Electrical and Electronic Engineering, Springer, [ISBN 978-981-15-4691-4], 2020, pp. 765-773.

[8] R.O.M. Team, How AI is Used in Today's Robots _ RIA Blog _ RIA Robotics Blog, 2018.

[9] A. Schroer, AI Robots_ 19 Examples Of Artificial Intelligence In Robotics _ Built In, 2019.
[10] D. Faggella, Machine Learning in Robotics – 5 Modern Applications, *Emerj*, 2019. <https://emerj.com/ai-sector-overviews/machine-learning-in-robotics/>. (accessed 05.12.20).
[11] A. Sharma, Artificial Intelligence Application in Robotics - AI Consulting Services, 2018.

CHAPTER EIGHT

Integrated deep learning for self-driving robotic cars

Tad Gonsalves[1] and Jaychand Upadhyay[2]
[1]Department of Information and Communication Sciences, Sophia University, Tokyo, Japan
[2]Xavier Institute of Engineering, Mumbai, India

8.1 Introduction

According to a recent survey, in the United States alone approximately 1.35 million people die in road crashes each year. On average 3700 people lose their lives every day on the roads. An additional 20–50 million suffer non-fatal injuries, often resulting in long-term disabilities [1]. A 2015 study by the US Department of Transportation NHTSA attributed 94% of accidents to human error, with only 2% to vehicular malfunction, 2% to environmental factors, and 2% to unknown causes [2]. Advanced autonomous driving or self-driving where humans are eliminated from the driving loop is considered to be the safest transportation technology of the future. Most major automobile manufacturers and IT enterprises worldwide are intensely engaged in developing a commercially viable prototype to seize the market-share [3]. In Japan, the government has set aside enormous funds to make autonomous driving technology a reality for the upcoming Olympics. The Society of Automotive Engineers International has proposed a conceptual model defining six levels of driving automation which range from fully manual level 0 to fully autonomous level 5. The model has become the de facto global standard adopted by stakeholders in the automated vehicle industry [4]. Although a fully functional level 5 self-driving car is still a long way off, most major automakers have reached advanced stages of developing self-driving cars [5,6].

The current autonomous vehicle (AV) technology is exorbitantly expensive as it tends to rely on highly specialized infrastructure and equipment like Light Detection and Ranging (LIDAR) for navigation, Global Positioning for localization, and Laser Range Finder for obstacle detection

Artificial Intelligence for Future Generation Robotics.
DOI: https://doi.org/10.1016/B978-0-323-85498-6.00010-1

[7,8] and so on. Some automakers have opted out of LIDARs because of their bulky size and exorbitant costs [9]. The focus of this study is to develop an agent that learns to drive by detecting road features and surroundings relying solely on camera vision.

A fully developed self-driving system needs countless experimental runs on public roads in all imaginable driving conditions. Autonomous driving cars have become a common site especially in some US cities and a few studies are found in the literature [10–14]. However, severe land restrictions and strict government laws do not permit firms in Japan to experiment autonomous driving on private and public roads. A viable solution is to create a fully functional but miniature driving circuit on a laboratory floor and thereby develop a self-driving prototype before launching it into a commercial product. We have developed such a self-driving machine learning environment in our laboratory. Through simulation and verification through a robocar we train and test individual self-driving agents to excel in the art of self-driving.

The self-driving software platform is built on the conceptual model of human driving which consists of four cyclic steps: perception, scene generation, planning, and action. The human driver takes in the traffic information through the senses, especially the vision and creates a 3D environment of the driving scenario around his/her car. Based on this scenario, the driver then plans the next action and finally executes the action via the steering, accelerator, brakes, indicators, and so on. The conscious cycle repeats itself throughout the course of driving. A self-driving software prototype can be built exactly along the same principles. It should be able to perceive the driving environment, generate the 3D scene, plan, and execute the action in rapidly iterating cycles.

Deep Learning algorithms, in general, perform supervised, unsupervised, semisupervised, and reinforcement learning (RL). The self-driving prototype integrates supervised and RL to achieve the autonomous driving goal. Perception and scene generation is achieved through supervised learning. Object detection and recognition algorithms break down the image picked up by the car cameras into individual components such as road signs, pedestrians, trees, buildings, traffic signals, vehicles, and so on. The isolated objects are then classified into as relevant or not relevant for driving. The relevant objects are then connected spatially to generate the 3D driving scene. The state-of-the-art technology used in image recognition is the convolutional neural network (CNN) [15] in various forms, such as AlexNet [16], VGGNet [17], and ResNet [18].

Deep Q Learning (DQL) is a subset of Deep Reinforced Learning. It combines CNN (to perceive the state of the environment in which the agent finds itself) and Deep Q Network (DQN) (to act on the perceived environment). Our driving agent (software program) learns to follow lanes and avoid randomly placed obstacles in the simulation environment using the DQL model. There is no dataset provided. Similarly, another agent learns to recognize traffic lights in the Laboratory for Intelligent and Safe Automobiles (LISA) dataset using the Faster R-CNN object recognition model. The agents are then integrated and loaded on robocar, which further continues learning on the laboratory floor.

Most of the car manufacturers and enterprises are engaged in performing self-driving experiments on private and public roads. This is possible if the government regulations allow self-driving vehicles on public roads. Traffic laws are strict in Japan making it very difficult for academia and industry to test self-driving vehicles on public roads. The only viable option is driving on private roads. Since land prices in Japan are exorbitantly high, most enterprises and universities do not venture into acquiring land and building infrastructure to deploy and test autonomous driving. What we propose in this study is a practical and viable option—design a miniature driving circuit on the laboratory floor and experiment with self-driving miniature robotic cars (Fig. 8.1). The laboratory driving circuit would contain the infrastructure essential for driving like roads, traffic signs, and signals. It will also contain peripheral objects like sidewalks,

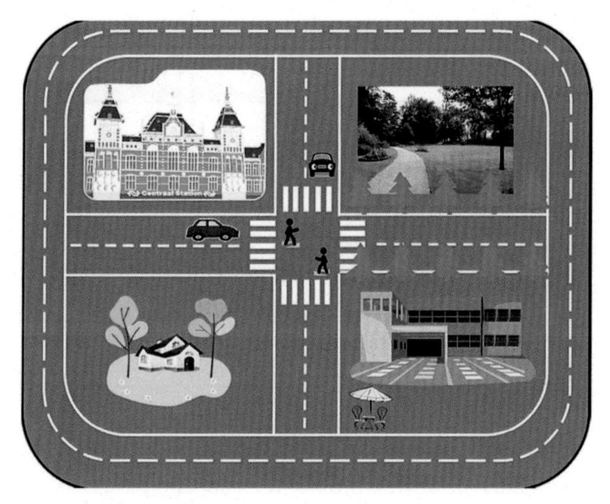

Figure 8.1 Laboratory floor driving circuit for robocar.

trees, buildings, pedestrians, etc., created by means of a 3D printer. The only difference from the real-life driving environment would be the scale. The fully developed self-driving prototype can then be deployed on public roads as a commercial product.

8.2 Self-driving program model

The self-driving program model is constructed on the human psychological autodriving cycle. The dynamic elements of the cycle and its imitation by autonomous driving cars is explained in the following subsections.

8.2.1 Human driving cycle

The self-driving program model described in this section is based on a psychological analysis of the human driving cycle which consists of the following four major steps: perception, scene generation, planning, and action (Fig. 8.2).

Perception

Perception refers to the taking in of all the information that impinges on our senses. Through our eyes, we perceive the signal and their colors, we read the traffic signs posted on the road, read the signs and words written on the road surface, and see the traffic moving in front of us. Mirrors

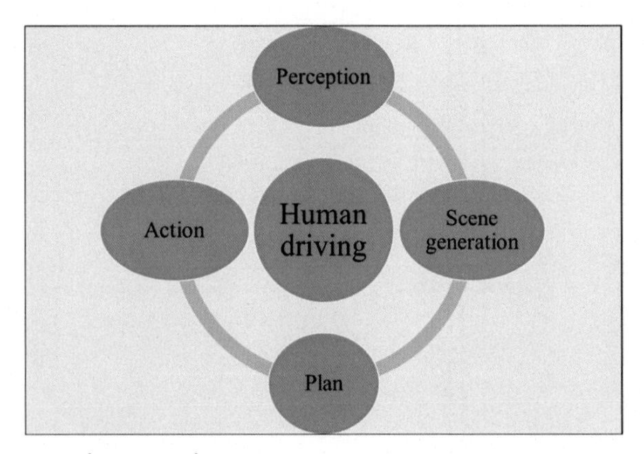

Figure 8.2 Human driving cycle.

fitted to the car and those by the roadsides act as secondary means of aiding our perception.

Scene generation

The information picked up by our sensory organs is transmitted to the brain via the nerves. The brain interprets the pieces of information supplied by each sense organ and combines them to build a world map of the surroundings in which the human driver's car happens to be at that point in time. From perception, humans generate the driving scene surrounding their vehicle. The mental image contains other vehicles, cyclists, pedestrians, signals, road signs, weather conditions, and so on.

Planning

The human driver knows from the basic rules of driving and from experience, how the situation will change the moment the signals controlling the flow of traffic through the crossing will change. The crossing will soon bustle with activity with traffic and pedestrians moving in all directions indicated by the green signals, while those patiently waiting as indicted by the red signals. With this in mind the driver plans his/her future actions.

Action

Finally, all the planning is put into action, triggered by the signal change. The human driver will release the brakes, gently press the gas pedal, and steer the vehicle. However, driving does not end after executing the planned action. The perceiving, scene generating, planning, and acting cycle repeats rapidly.

8.2.2 Integration of supervised learning and reinforcement learning

Autonomous driving is achieved through a combination of two phases of learning described in the following subsections:

Supervised learning

Supervised learning takes place aided by a supervisor that guides the learning agent. The learning agent is the machine learning (ML) algorithm or model and the supervisor is the output in the data for a given set of inputs. The aim of the learning algorithm is to predict how a given set of inputs leads to the output. At first, the ML agent takes the inputs and randomly predicts the corresponding outputs. Since the random calculation is akin to shooting in the dark, the predicted outcomes are far away from the known outcomes. The supervisor at the output end indicates the error in prediction which again guides the learning agent to minimize the error.

The first two steps, namely, perception and scene generation in autonomous driving, are performed through supervised learning. The program is trained using a large number of digital photographs involving driving scenes. In the learning phase, the program identifies and classifies the various objects in the photo images. The master program in the AV controller computer calls the respective ML trained subroutines and combines their result to generate the driving scene surrounding the AV. This act corresponds to sensor readings fusion in real-life self-driving vehicles that use an array of sensors to generate a mapping of the surroundings.

Reinforcement learning

Loosely speaking, RL is very similar to teaching a dog learn new tricks. When the dog performs the trick as directed, it is rewarded; when it makes mistakes, it is corrected. The dog soon learns the tricks by mastering the policy of maximizing its rewards at the end of the training session. Technically, RL algorithms are used to train biped robots to walk without hitting obstacles, self-driving cars to drive by observing traffic rules and avoiding accidents, software programs to play games against human champions and win. The agent (software program engaged in learning) is not given a set of instructions to deal with every kind of situation it may encounter as it interacts with its environment. Instead, it is made to learn the art of responding correctly to the changing environment at every instant of time. RL consists of a series of state and actions. The agent is placed in an environment which is in some state at some time. When the agent performs an action, it changes the state of the environment which evaluates the action and immediately rewards or punishes the agent. A typical RL algorithm tries mapping observed states onto actions, while optimizing a short-term or a long-term reward. Interacting with the environment continuously, the RL agent tries to learn policies, which are sequences of actions that maximize the total reward. The policy learned is usually an optimal balance between the immediate and the delayed rewards [19]. There are a variety of RL algorithms and models found in literature [20]. This study concentrates on the use of Deep-Q learning, first to train the self-driving agent in the simulation environment and then fine-tune the robocar driving on the laboratory floor.

8.3 Self-driving algorithm

Driving functions are divided into three major categories as shown in Fig. 8.3:

- Fundamental driving functions: these include functions like white lane detection, traffic signals detection, traffic signs recognition, and driving on paved roads that do not have lane markings (laneless driving).
- Hazards detection like obstacle detection, collision detection, and risk level estimation.
- Warning systems like driver monitoring, pedestrian hazard detection, and sidewalk cyclists detection.

These categories of functions are seamlessly integrated in the autonomous driving prototype. The individual functions and their respective training are explained below.

8.3.1 Fundamental driving functions

White lane detection

Semantic segmentation technique is used to detect white lanes painted on the surface of the roads. VCG net is used in training and testing of white lane datasets. Fig. 8.4 shows the results obtained in detecting white lanes.

In Fig. 8.4, several white lanes are visible. There are also white metallic fences along the road signs which demark pedestrian zones. These lanes

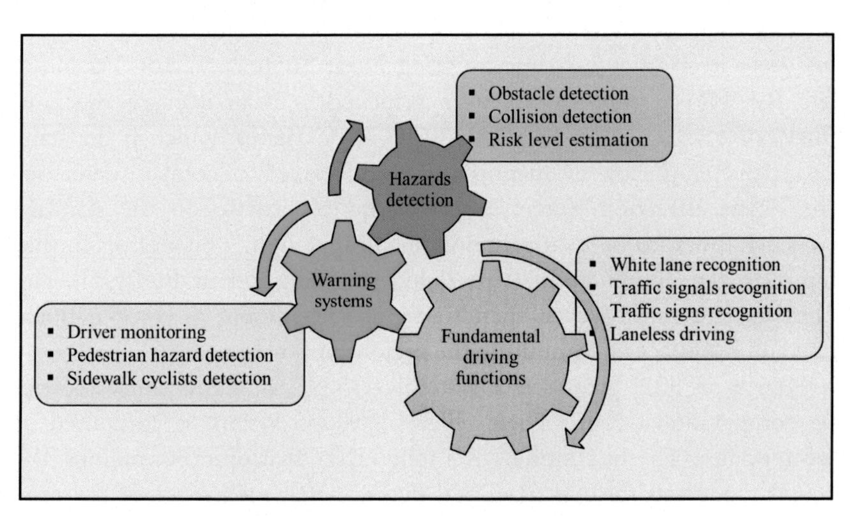

Figure 8.3 Three integrated modules that constitute autonomous driving.

Figure 8.4 White lane detection.

are colored in orange by the semantic segmentation learning network. As can be seen from the bottom diagram, the trained program does not color the white fences. In other words, it has learnt to clearly distinguish between the white lanes and the rest of the lane-like white objects.

Signals

Ross Girshick et al. proposed a Regions with CNN (R-CNN) approach for object detection in a given image [21]. This system takes an input image, extracts a large number of bottom-up region proposals (in the order of 2k), computes features for each proposal using a CNN. In the final step, it classifies each region using linear Support Vector Machines. The R-CNN is reported to have achieved a mean average precision (mAP) of 43.5% on the PASCAL VOC 2010 dataset, which is the standard benchmark for evaluating the performance of object detectors. The object detection system, although quite accurate on the standard dataset, is found to be relatively slow in computation. The authors further improved the computational speed by proposing a Fast R-CNN. The improved system, however, spends a significant amount of computational time and resources in computing the region proposals.

Ren et al. [22] proposed a Faster R-CNN that significantly reduces the computational time. Their object detection system is composed of two modules. The first module is a fully CNN that proposes regions; the second module is the Fast R-CNN detector that makes use of the proposed regions. Functioning as a single, unified network for object

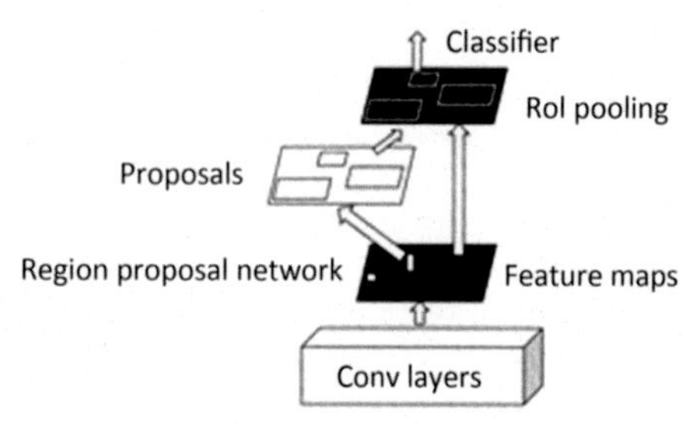

Figure 8.5 Region proposal network.

detection, the Region Proposal Network (RPN) module tells the Fast R-CNN module where to look for objects in the image (Fig. 8.5). By sharing convolutions at test-time, the marginal cost for computing proposals is found to be small (approximately 10 ms per image).

We use a variation of the Faster R-CNN for traffic signal recognition. The images are fed to the convolutional layers which give us the feature maps. These are in turn fed to the region proposal network that outputs region proposals. The region proposal along with the "Region of Interest" (RoI) obtained through the RoI pooling are finally fed into the classifier. The classifier classifies the input image as well as its relative location by means of a bounding box called the anchor.

8.3.2 Signals

Fig. 8.6 shows six classes of traffic lights in the LISA dataset. The average prediction (AP) for each of the classes is shown in Table 8.1. The first column shows the different classes of the signals in the LISA dataset. It should be noted that the classification is not just red, yellow, and green signals, but detailed versions of the signals as they are found in city areas with dense traffic. The next two columns depict the number of instances in the training and the test sets. The final column shows the AP accuracy of each class. This prediction accuracy is more than 92% in all the classes except the "goleft" class. The lower accuracy is due to the presence of noise in the dataset (Fig. 8.7).

Traffic signs

The traffic signs posted along the sidewalks are picked up by the front camera and recognized in real-time by the deep learning network. To improve the learning accuracy of the traffic signs recognition system, learning is divided into two phases. In the first phase, signs that are clearly

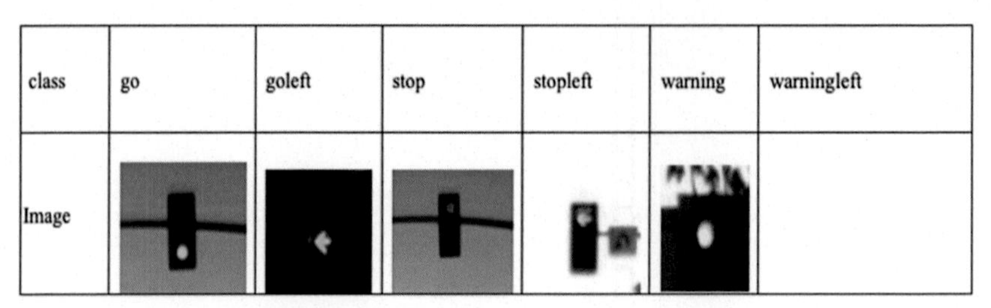

class	go	goleft	stop	stopleft	warning	warningleft
Image						

Figure 8.6 Six classes of traffic lights in the LISA dataset.

Table 8.1 Train and test results of traffic signal recognition.

Classes	Training	Test	Average prediction
go	9053	1141	92.40%
goleft	425	228	62.49%
stop	10621	1482	96.99%
stopleft	4172	316	94.45%
warning	318	146	99.63%
warningleft	90	109	92.57%

Figure 8.7 An instance of signal classes recognized by the trained agent.

visible in the camera are used for training. The recognition results produced by the algorithm are shown in Fig. 8.8. However, just learning to recognize clearly visible road signs is not sufficient for safe driving in real life. The second phase consists of learning to recognize faded and not clearly

Figure 8.8 Detection of clearly visible traffic signs.

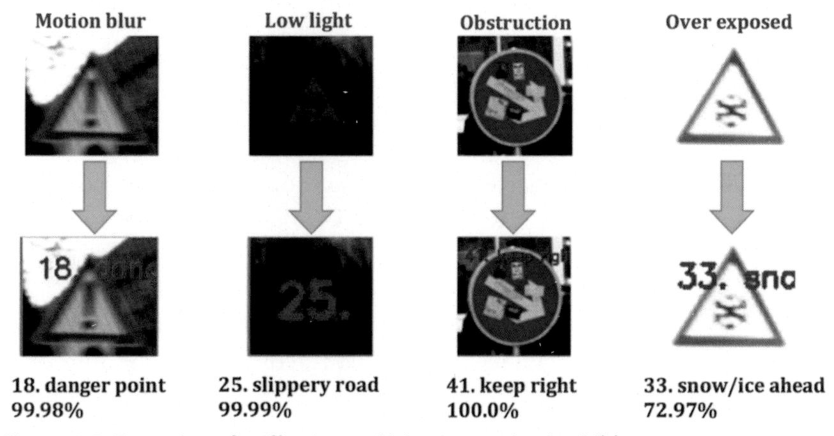

Figure 8.9 Detection of traffic signs which are not clearly visible.

visible road signs. The test results are shown in Fig. 8.9. Learning and extensive testing are carried out using the German Traffic Sign Dataset [23] containing 50,000 images belonging to 45 classes, and the Belgium Traffic Sign Dataset [24], which includes more than 7000 single-images of traffic signs found in Belgium divided into 62 classes.

Laneless driving

It is easier to drive in cities and in other urban areas where the lanes are clearly marked on the road. The AV has to detect the nearest lane to its left and to its right and steer forward trying to keep itself in the middle of the two bounding lanes at all times. However, driving in the mountains or the countryside, even if the roads are paved, is a major challenge.

There is no reference point for the vehicle to steer by. Although the other modules like obstacle detection, traffic signals and signs detection, pedestrian detection, and so on will function as in the urban setting, helping the AV to keep away from danger and accidents, there will not be any clear guidelines for the vehicle steering.

The laneless driving module deals with the recognition of the shoulder of the paved road, even if there are no clear markings. Sometimes, the edges are broken, at other times they are not clearly delineated. Semantic segmentation detects and segments the surface area of the road around the car which is safe for driving (Fig. 8.10A and B). The AV positions itself at the center of the two lateral edges and steers forward.

8.3.3 Hazards

YOLO and detection of objects

You Only Look Once (YOLO) v3 [25,26] is trained on the Open Images Dataset v4 which has 15,440,132 bounding boxes arranged in 600 categories of objects. The salient feature of YOLO is its speed of recognition of objects. The objects are placed into bounding boxes and then recognized as belonging to a particular class (Fig. 8.11).

Collision avoidance

Collision avoidance is one of the principal components of safe driving. Various scenarios and techniques for collision avoidance in the case of a single AV driving are found in the literature: integrated trajectory control [27], fuzzy danger level detection [28], side collision avoidance [29], collision avoidance at traffic intersections [30], collision avoidance, and vehicle stabilization [31]. However, currently AI collision avoidance techniques are mechanical and brittle; they do not take into consideration the human-way of driving and responding to uncertain situations. In our study, we go a step further and make our agent human-like in avoiding collisions.

Humans do not wait until the last minute to avoid collisions, as shown in Fig. 8.12A. Relying on the visual information at hand, they try to avoid colliding with obstacles when viewed from far, as shown in Fig. 8.12B. The autonomous navigation system for obstacle avoidance in dynamic driving environments learns the art of maneuvering through roads strewn with moving obstacles using the Q-learning algorithm.

The above human-like collision avoidance system (Fig. 8.12) is further trained to drive in an extreme situation involving two obstacles with varying priority levels of collision avoidance. The Q-learning algorithm

Figure 8.10 Driving on paved roads without lane markings: (A) paved road without lanes; (B) unpaved road without lanes; (C) segmented paved road; and (D) segmented unpaved road.

rewards the agent with greater score to avoid colliding with a high priority (high risk) object than with the low priority (low risk) object. The agent, accordingly, learns to avoid the higher risk obstacle when faced with two hazardous obstacles with varying priorities. The following two kinds of human-like behaviors are demonstrated by the AI navigation system:

1. When the two moving obstacles do not come within the danger zone, the AINS passes smoothly in-between the two keeping a fairly safe distance from them (Fig. 8.13A).

Figure 8.11 Bounding boxes are placed by You Only Look Once on individual objects detected.

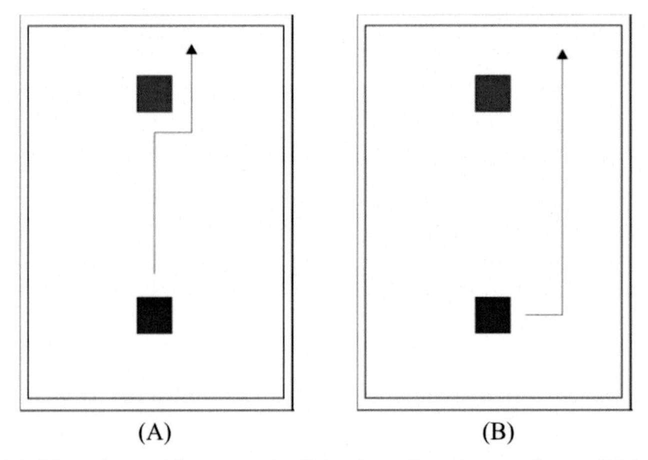

Figure 8.12 Obstacle avoidance types: (A) naïve obstacle avoidance; (B) human-like obstacle avoidance.

2. When the two moving obstacles come within the danger zone, the AINS chooses to avoid the high-priority obstacle at all costs even though it may near-miss hitting the low-priority obstacle (Fig. 8.13B).

Estimation of risk level for self-driving

As a self-driving vehicle keeps on driving, following the various traffic rules like lanes, signals, traffic signs, it has to be aware of the other traffic

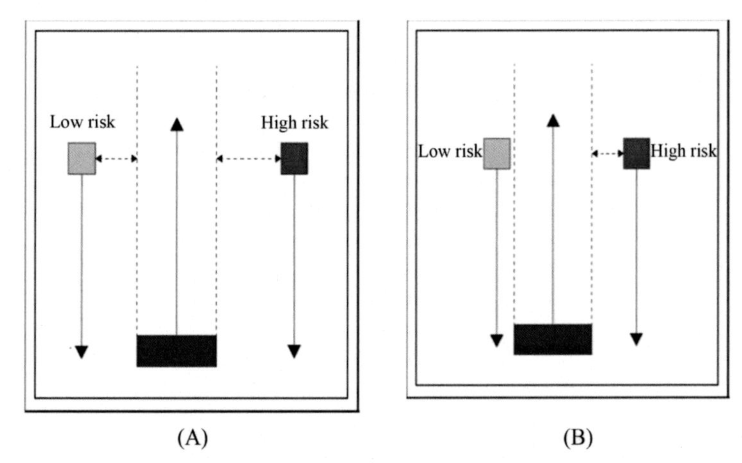

Figure 8.13 (A) Avoiding low and high priority obstacles. (B) Avoiding near miss collisions.

around it at any given point in time. Pedestrians, motorcyclists, and cars often come very close to an AV. This section deals with the real-time computation of the risk level of nearby objects to an AV. The objects are given a priori priority, like humans, cars, motorbikes, animals, and leaves, which can be detected by the front, side, and back mirrors. The risk level is computed using fuzzy logic and the Genetic Algorithm (GA) [32].

The fuzzy rules are framed as follows:

Rule1: If the priority is low and the distance is far, then the risk level is 1.

Rule2: If the priority is high and the distance is far, then the risk level is 5.

Rule3: If the priority is low and the distance is close, then the risk level is 6.

Rule4: If the priority is high and the distance is close, then the risk level is 10.

The membership functions shown in Fig. 8.14 are standard triangular functions. The shapes of thee membership functions are optimized to suit the risk calculator using the GA. The overall system resulting from the combination of fuzzy logic and GA is shown in Fig. 8.15. The inputs to the system are priority and distance of a traffic object (vehicle, human, animal, leaves). The resultant of the GA fuzzy controller is the risk level in the range 0–1.

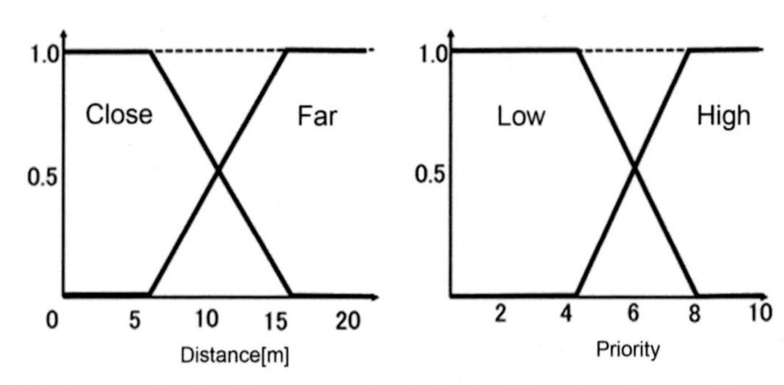

Figure 8.14 Fuzzy rules for distance and priority in risk estimation model.

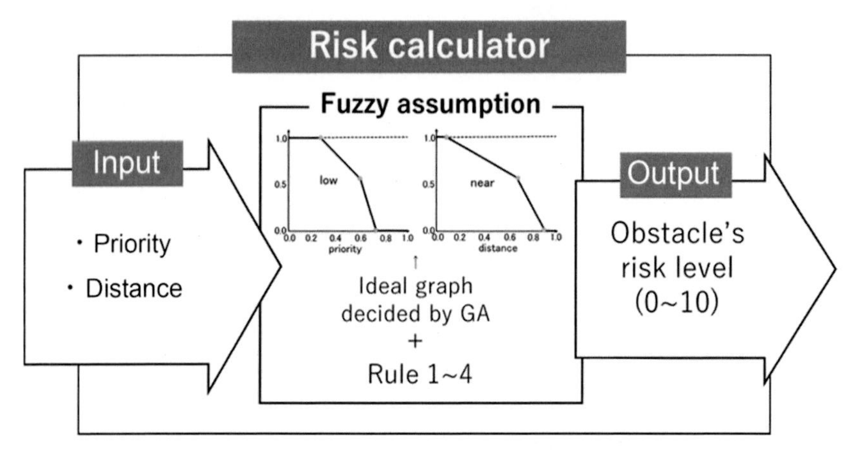

Figure 8.15 Risk computation system using fuzzy logic and Genetic Algorithm.

8.3.4 Warning systems

Driver monitoring

Feeling drowsy or falling asleep at the steering wheel is one of the most frequent causes of car accidents and crashes. The driver-monitoring system takes real-time pictures of the driver's face and feeds them into the trained VCG net [17]. The trained network determines if the driver is showing signs of drowsiness and sleepiness (Fig. 8.16).

Pedestrian hazard detection

Caltech Pedestrian Detection Benchmark [33,34] which is a 640 × 480, 30 Hz, 10 hour video containing 2300 pedestrians enclosed in 350,000 bounding boxes is used for training and testing. The DL program is a combination of YOLO and SE-ResNet. The hazard levels are computed

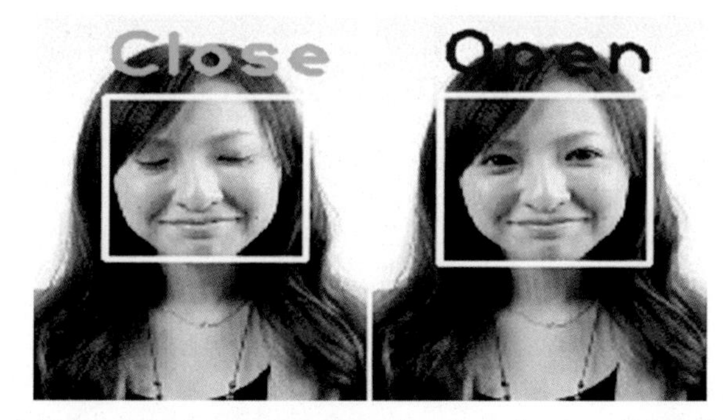

Figure 8.16 Driver monitoring system detecting drowsiness and sleepiness.

Figure 8.17 Pedestrian hazard levels: (A) pedestrian hazard level = 0.2833; (B) pedestrian hazard level = 0.0436; (C) pedestrian hazard level = 0.6644; (D) pedestrian hazard level = 0.3044.

taking into consideration not just the distance of the pedestrians from the self-driving vehicle, but their orientation and speed of walking as well. For example, Fig. 8.17A and B shows two pedestrians exactly at the same distance from the AV. However, pedestrian (A) is stationary and is about to cross the road, while pedestrian (B) is briskly walking away from the vehicle on the sidewalk. Similarly, the hazard level of the pedestrian who is about to cross the road (C) is much higher than the one who is walking away from the road after crossing (D).

Sidewalk cyclists' detection

Moving object detection is one of the fundamental technologies necessary to realize autonomous driving. Object detection has received much attention in recent years and many methods have been proposed (Fig. 8.18).

However, most of them are not enough to realize moving object detection, chiefly because these technologies perform object detection from one single image. The future projected position of a moving object is learnt from its past and current positions. In this study, we use Deep Convolutional Generative Adversarial Networks (DCGANs) [35–40] to predict the future position of cyclists riding along the sidewalks. They pose a threat to vehicles on the road and suddenly jutting onto the road from the sides. Monitoring their movements and predicting their behavior is a great help in overcoming accidents.

8.4 Deep reinforcement learning

8.4.1 Deep Q learning

Q learning is a variation of RL. In Q learning, there is an agent with states and corresponding actions. At any moment, the agent is in some feasible state. In the next time step, the state is transformed to other state(s) by performing some action. This action is accompanied either by reward or a punishment. The goal of the agent is to maximize the reward

Figure 8.18 Predicting the behavior of a cyclist crossing the road in future time.

gain. The Q learning algorithm is represented by the following update formula [39]:

$$Q(st,\ at) \leftarrow Q(st,\ at) + \alpha(rt - Q(st,\ at) + \gamma Q(st + 1,\ a')) \tag{8.1}$$

where Q(st, at) represents the Q value of the agent in the state st, and action at time t, rewarded with reward rt., α is the learning rate and γ is the discount factor. The γ parameter is in the range [0,1]. If γ is closer to 0, the agent will tend to consider only immediate rewards. On the other hand, if it is closer to 1, the agent will consider future rewards with greater weight, and thus is willing to delay the reward. The Learning rate and Discount factor, described below, are the two most crucial parameters influencing the performances of the Q learning algorithm.

Learning rate

The learning rate determines the strength with which the newly acquired information will override the old information. A factor of 0 will make the agent not learn anything, while a factor of 1 will make the agent consider only the most recent information. In fully deterministic environments, the learning rate is optimal. When the problem is stochastic, the algorithm still converges under some technical conditions on the learning rate, that requires it to decrease to zero. In practice, often a constant learning rate is used.

Discount factor

The discount factor determines the importance of future rewards. A factor of 0 will make the agent short-sighted by only considering current rewards, while a factor approaching 1 will make it strive for a long-term high reward. If the discount factor exceeds 1, the action values may diverge. Even with a discount factor only slightly lower than 1, the Q learning leads to propagation of errors and instabilities when the value function is approximated with an artificial neural network. Starting with a lower discount factor and increasing it toward its final value yields accelerated learning.

8.4.2 Deep Q Network

CNN is essentially a classification structure for classifying images into labeled classes. The various layers of the CNN extract image features and finally learn to classify the images. Hence, the outputs of a typical CNN represent the classes or the labels of the classes, the CNN has learnt to classify (Fig. 8.19). A DQN is a variation of CNN. The outputs are not classes, but the Q values (or rewards) corresponding to each of the actions the agent has learnt to take in response to its state in the environment. In our model, the input to the DQN is the image of the street the car sees in front of it at a given point of time. The output is the steering angle.

8.4.3 Deep Q Network experimental results

The action plan consists of the number of actions the agent can take for a given environment state captured by the front camera. In all, we trained four different agents, each with a different action plan. Table 8.2 shows the specification of each DQN model with its corresponding action plan (steering angle) (Fig. 8.20).

Each model has been trained for 1,000,000 cycles. The graph in Fig. 8.21 shows the simple moving averages of the score (reward) for time intervals of 500 episodes.

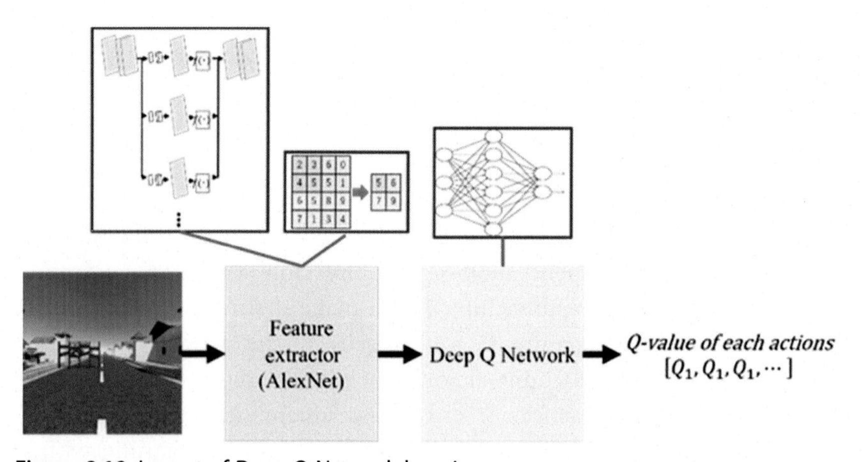

Figure 8.19 Layout of Deep Q Network learning.

Table 8.2 Actions performed by the Deep Q agent.

Model	Number of actions	Action list (steering angles)
1	3	[−25, 0, 25]
2	5	[−25, −20, 0, 20, 25]
3	7	[−30, −20, −10, 0, 10, 20, 30]
4	7	[−25, −20, −10, 0, 10, 20, 30]

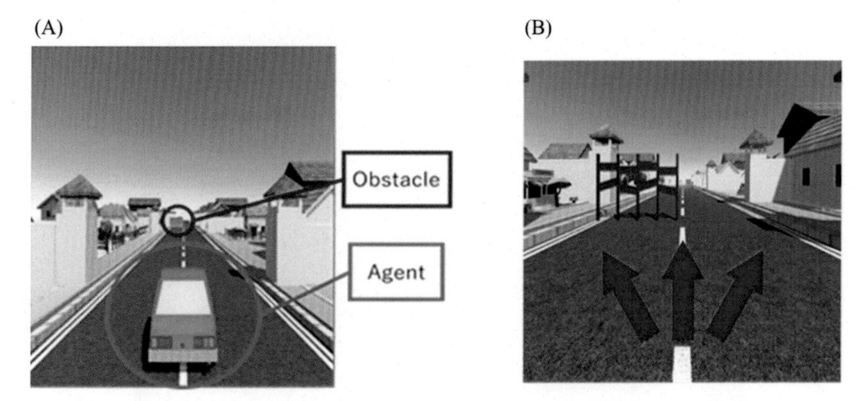

Figure 8.20 (A) Agent keeping lanes and avoiding obstacles (left). (B) Steering actions taken by agent (right).

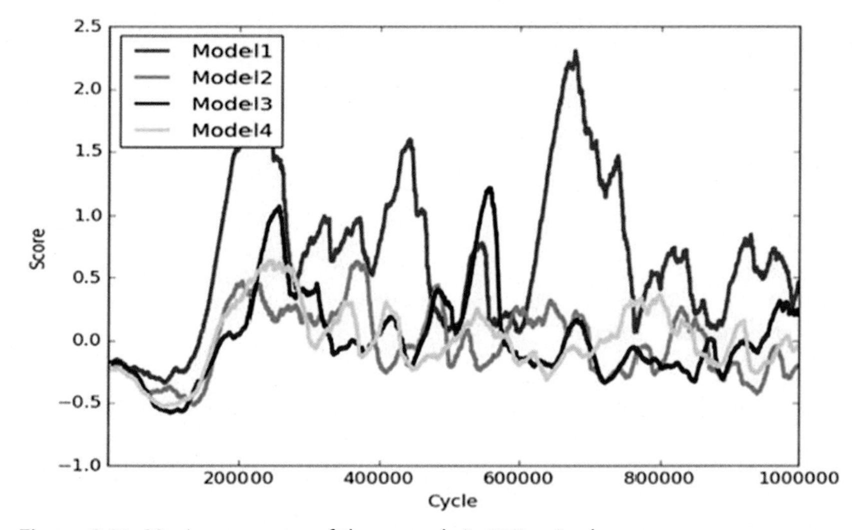

Figure 8.21 Moving averages of the rewards in 500 episodes.

8.4.4 Verification using robocar

The laboratory driving circuit is shown in Fig. 8.10. The robotic car, called "Robocar 1/10" is 1/10th the size of the real car with all the real-life driving functions, manufactured by ZMP Inc., Japan (https://www.zmp.co.jp/en/products/robocar-110).

Two monovisual cameras are fitted on the car. The lower camera is dedicated to capturing the lanes and obstacles on the road and sidewalks, while the upper camera captures traffic lights and signs. The two separate

Figure 8.22 Laboratory verification.

program agents—lane and obstacles, and traffic lights detecting modules—are integrated in the hard disk of the Unix running Robocar. The test runs verify the extent to which the agents have learned to drive along the lanes, avoiding obstacles and recognizing and thereby obeying the signals (Fig. 8.22).

8.5 Conclusion

Current state-of-the-art autonomous driving is an extremely expensive technology because it relies heavily on expensive equipment and road infrastructure. In this study, we have presented an integrated ML approach based only on visual information obtained by the cameras mounted in the front of the car. The integrated approach consists of supervised learning for recognizing traffic objects like signals, road signs, obstacles, pedestrians, etc., and deep RL for driving safely observing the traffic rules and regulations.

The self-driving software platform is built on the conceptual model of human driving which consists of four cyclic steps: perception, scene generation, planning, and action. The human driver takes in the traffic information through the senses, especially the vision and creates a 3D environment of the driving scenario around his/her car. Based on this scenario, the driver then plans the next action and finally executes the action via the steering, accelerator, brakes, indicators, and so on. The conscious cycle repeats itself throughout the course of driving. A self-driving software prototype can be built exactly along the same principles. It can perceive the driving environment, generate the 3D scene, plan, and execute the action in rapidly iterating cycles.

The software agents trained in the simulation environment are embedded in Robocar 1/10 which verifies the self-driving learnt. It is further trained by manually operating a remote set of steering, acceleration, and brake controls. The end result is a fully trained self-driving prototype without providing expansive land with private or public roads embedded with specialized infrastructure.

In our future study, we plan to integrate all the driving modules (agents) into a unified entity and mount it on a computer installed on a car which can drive on private and public roads. The low-cost agent will further learn to assess real-life driving situations and be able to deliver safe drive.

References

[1] https://www.asirt.org/safe-travel/road-safety-facts/
[2] National Highway Traffic Safety Administration, Critical reasons for crashes investigated in the national motor vehicle crash causation survey, Traffic Safety Facts, DOT HS 812 115, February, 2015. <https://crashstats.nhtsa.dot.gov/Api/Public/ViewPublication/812115>
[3] M. Aly, Real time detection of lane markers in urban streets, 2008 IEEE Intelligent Vehicles Symposium, Eindhoven, 2008, pp. 7−12.
[4] https://www.synopsys.com/automotive/autonomous-driving-levels.html
[5] M. Bojarski, D. Del Testa, D. Dworakowski, B. Firner, B. Flepp, P. Goyal, et al. (2016), End to End Learning for Self-Driving Cars, arXiv:1604.07316.
[6] B. Lawrence, S. Christopher, Autonomy: The Quest to Build the Driverless Car—And How It Will Reshape Our World, Ecco, Reprint Edition, August 28, 2018.
[7] N. Buch, S.A. Velastin, J. Orwell, A review of computer vision techniques for the analysis of urban traffic, IEEE Trans. Intell. Transp. Syst. 12 (3) (2011) 920−939. Sept.
[8] Z. Chen, X. Huang, End-to-end learning for lane keeping of self-driving cars, 2017 IEEE Intelligent Vehicles Symposium (IV), Los Angeles, CA, 2017, pp. 1856−1860.
[9] H. McKenzie, Insane Mode: How Elon Musk's Tesla Sparked an Electric Revolution to End the Age of Oil. Dutton, November 27, 2018.
[10] L. Chi, Y. Mu, Deep Steering: Learning End-to-End Driving Model from Spatial and Temporal Visual Cues. arXiv:1708.03798v1 [cs.CV] 12 Aug 2017.
[11] J. Condliffe, Lidar Just Got Way Better—But It's Still Too Expensive for Your Car, MIT Technology Review, November 28, 2017. Retrieved from <https://www.technologyreview.com/s/609526/lidar-just-got-way-better-but-its-still-too-expensive-for-your-car/>
[12] M.S. Darms, P.E. Rybski, C. Baker, C. Urmson, Obstacle detection and tracking for the urban challenge, IEEE Trans. Intell. Transp. Syst. 10 (3) (2009) 475−485. Sept.
[13] J. Funke, M. Brown, S.M. Erlien, J.C. Gerdes, Collision Avoidance and stabilization for autonomous vehicles in emergency scenarios, IEEE Trans. Control Syst. Technol. 25 (4) (2017) 1204−1216. July.

[14] R. Girshick, J. Donahue, T. Darrell, J. Malik, Rich feature hierarchies for accurate object detection and semantic segmentation, in: IEEE Conference on Computer Vision and Pattern Recognition (CVPR), 2017.
[15] C. Szegedy, W. Liu, Y. Jia, P. Sermanet, S. Reed, D. Anguelov, et al., Going deeper with convolutions, in: Proceedings of the IEEE Conference on Computer Vision and Pattern Recognition, 2015, pp. 1−9.
[16] A. Krizhevsky, I. Sutskever, G.E. Hinton, ImageNet classification with deep convolutional neural networks, Adv. Neural Inf. Process. Syst. (2012) 1097−1105.
[17] K. Simonyan, A. Zisserman, Very Deep Convolutional Networks for Large-Scale Image Recognition, 2014, arXiv:1409.1556.
[18] K. He, X. Zhang, S. Ren, and J. Sun, Deep residual learning for image recognition, in: Proceedings of the IEEE conference on computer vision and pattern recognition, 2016, pp. 770−778.
[19] R.S. Sutton, A.G. Barto, Reinforcement Learning: An Introduction (Adaptive Computation and Machine Learning series), A Bradford Book; second edition, November 13, 2018.
[20] M. Lapan, Deep Reinforcement Learning Hands-On: Apply modern RL Methods, with Deep Q-Networks, Value Iteration, Policy Gradients, TRPO, AlphaGo Zero and More, Packt Publishing, 2018.
[21] R. Girshick, Fast R-CNN, in IEEE International Conference on Computer Vision (ICCV), 2015.
[22] S. Ren, K. He, R. Girshick, J. Sun, "Faster R-CNN: towards real-time object detection with region proposal networks, IEEE Trans. Pattern Anal. Mach. Intell. 39 (6) (2017) 1137−1149. June 1 2017.
[23] J. Stallkamp, M. Schlipsing, J. Salmen, and C. Igel, The German traffic sign recognition benchmark: a multi-class classification competition, in Proceedings of the IEEE International Joint Conference on Neural Networks, 2011, pp. 1453−1460.
[24] R. Timofte, K. Zimmermann, L. Van Gool, Multi-view traffic sign detection, recognition, and 3D localisation, Mach. Vis. Appl. 25 (3) (2014) 633−647.
[25] J. Redmon, A. Farhadi, Yolov3: an incremental improvement, 2018. arXiv:1804.02767.
[26] B. Benjdira, T. Khursheed, A. Koubaa, A. Ammar, and K. Ouni, Car detection using unmanned aerial vehicles: Comparison between faster r-cnn and yolov3, in: 2019 1st International Conference on Unmanned Vehicle Systems-Oman (UVS), IEEE, February 2019, pp. 1−6.
[27] J. Verhaegh, J. Ploeg, E. van Nunen, A. Teerhuis, Integrated trajectory control and collision avoidance for automated driving, 2017 5th IEEE International Conference on Models and Technologies for Intelligent Transportation Systems (MT-ITS), Naples, Italy, 2017, pp. 116−121.
[28] Z. Ramyar, S. Salaken, S.M. Homaifar, A. Nahavandi, A. Karimoddini, A collision avoidance system with fuzzy danger level detection, 2017 IEEE Intelligent Vehicles Symposium (IV), Los Angeles, CA, USA, 2017, pp. 283−288.
[29] Y.K. Ou, C.W. Fu, Y.L. Chen, In-vehicle human interface design for side collision avoidance system, 2017 International Conference on Applied System Innovation (ICASI), Sapporo, 2017, pp. 542−545.
[30] H. Ahn, D. Del Vecchio, Safety verification and control for collision avoidance at road intersections, IEEE Trans. Autom. Control. PP (99) (2016) pp. 1−1.
[31] V.M. Milan Kumar, Shenbagaraman, A. Ghosh, Predictive data analysis for energy management of a smart factory leading to sustainability. Book Chapter [ISBN 978-981-15-4691-4] in: M.N. Favorskaya, S. Mekhilef, R.K. Pandey, N. Singh (Eds.), Innovations in Electrical and Electronic Engineering, Springer, 2020, pp. 765−773.

[32] J. Funke, M. Brown, S.M. Erlien, J.C. Gerdes, Collision avoidance and stabilization for autonomous vehicles in emergency scenarios, IEEE Trans. Control. Syst. Technol. 25 (4) (2017) 1204–1216. July 2017.
[33] T. Gonsalves, Artificial Intelligence: A Non-Technical Intelligence, Sophia University Press, Tokyo, 2017.
[34] M. Sampurna, E.B. Valentina, N.S. Rabindra, G. Ankush, Prediction analysis of idiopathic pulmonary fibrosis progression from OSIC dataset, 2020 IEEE International Conference on Computing, Power and Communication Technologies (GUCON), 2–4 Oct. 2020, pp. 861–865, Available from https://doi.org/10.1109/GUCON48875.2020.9231239
[35] M. Andriluka, S. Roth, B. Schiele. People-tracking-by-detection and people-detection-by-tracking. In CVPR, 2008.
[36] S. Mandal, S. Biswas, V.E. Balas, R.N. Shaw, A. Ghosh, Motion prediction for autonomous vehicles from Lyft dataset using deep learning, 2020 IEEE 5th International Conference on Computing Communication and Automation (ICCCA) 30–31 Oct. 2020, pp. 768–773, Available from https://doi.org/10.1109/ICCCA49541.2020.925079
[37] P. Dollár, C. Wojek, B. Schiele, P. Perona, Pedestrian detection: a benchmark, in: 2009 IEEE Conference on Computer Vision and Pattern Recognition, IEEE, June 2009, pp. 304–311.
[38] Y. Belkhier, A. Achour, R.N. Shaw, Fuzzy passivity-based voltage controller strategy of grid-connected PMSG-based wind renewable energy system, 2020 IEEE 5th International Conference on Computing Communication and Automation (ICCCA), Greater Noida, India, 2020, pp. 210–214. Available from https://doi.org/10.1109/ICCCA49541.2020.9250838
[39] I.J. Goodfellow, J. Pouget-Abadie, M. Mirza, B. Xu, D. Warde-Farley, S. Ozair, et al., Generative adversarial networks, NIPS, 2014.
[40] A. Radford, L. Metz, S. Soumith, Un-supervised Representation Learning With Deep Convolutional Generating Adversarial Networks, CoRR, 2015.

Further reading

J. Fan, Z. Wang, Y. Xie, Z. Yang, A theoretical analysis of deep Q-learning, Learning for Dynamics and Control, 2020, pp. 486–489. PMLR.
C. Innocenti, H. Lindén, G. Panahandeh, L. Svensson, N. Mohammadiha, Imitation learning for vision-based lane keeping assistance, 2017 IEEE 20th International Conference on Intelligent Transportation Systems (ITSC), Yokohama, 2017, pp. 425–430.
P. Isola, J. Zhu, T. Zhou, A.A. Efros, Image-to-image translation with conditional adversarial networks, CVPR, 2017.
M. B. Jensen, M. P. Philipsen, A. Møgelmose, T. B. Moeslund, M.M. Trivedi, Vision for looking at traffic lights: issues, survey, and perspectives IEEE Trans. Intell. Transp. Syst., vol PP(99) 1–16.
J. Ji, A. Khajepour, W.W. Melek, Y. Huang, Path planning and tracking for vehicle collision avoidance based on model predictive control with multiconstraints, IEEE Trans. Veh. Technol. 66 (2) (2017) 952–964. Feb. 2017.
V. John, K. Yoneda, B. Qi, Z. Liu, S. Mita, Traffic light recognition in varying illumination using deep learning and saliency map, 17th International IEEE Conference on Intelligent Transportation Systems (ITSC), Qingdao, 2014, pp. 2286–2291.
D.H. Kang et al., Vision-based autonomous indoor valet parking system, 2017 17th International Conference on Control, Automation and Systems (ICCAS), Jeju, 2017, pp. 41–46.

B. O'Donoghue, I. Osband, R. Munos, V. Mnih, The uncertainty bellman equation and exploration, International Conference on Machine Learning, 2018, pp. 3836–3845.

I. Sorokin, A. Seleznev, M. Pavlov, A. Fedorov, A. Ignateva, Deep Attention Recurrent Q-Network, 2015. arXiv:1512.01693.

CHAPTER NINE

Lyft 3D object detection for autonomous vehicles

Sampurna Mandal[1], Swagatam Biswas[1], Valentina E. Balas[2], Rabindra Nath Shaw[3] and Ankush Ghosh[1]

[1]School of Engineering and Applied Sciences, The Neotia University, Kolkata, India
[2]Department of Automatics and Applied Software, Aurel Vlaicu University of Arad, Arad, Romania
[3]Department of Electrical, Electronics & Communication Engineering, Galgotias University, Greater Noida, India

9.1 Introduction

Living in this 21st century has provided us with many rare opportunities and technologies to implement and that can improve the quality of our daily lives. One such technology is self-driving cars aka, automated vehicles (AV). Avoidable collisions, single-occupant commuters, vehicle emissions, etc. are choking cities, while infrastructure is growing rapidly. Autonomous vehicles are expected to come into this situation and vastly improve the current conditions and redefine transportation by unlocking a myriad of societal, environmental, and economic benefits. But at the same time, making a fully functional automated vehicle is not an easy task by any stretch. The technical research required to develop higher level autonomy functions like perception, prediction, and planning is extremely high and mostly out of the reach of the "normal" public. However, this dataset provided by Lyft aims to democratize access to such data, so that developers like us can work on it from our homes, and hopefully make some progress toward building the perfect AV.

Driving a motorized vehicle is not a simple task, as the drivers have to take into consideration many things (cars, cyclists, pedestrians, other movable objects, and traffic lights signals) before moving themselves. All this information can be a lot for a new driver, or can cause panic for anyone when put in a dire situation; this is when mistakes happen, resulting in accidents [1].

Efforts have been made to make driving as safe as possible by adding smart features, such as proximity sensors, LiDAR, and cameras, to cars to

Artificial Intelligence for Future Generation Robotics.
DOI: https://doi.org/10.1016/B978-0-323-85498-6.00003-4

avoid such accidents. But these alone are never enough. Nowadays, with exceptional leaps in artificial intelligence and the computational capabilities of computers, powerful computers can be inserted into a car which take inputs from all the sensors and cameras around the car to gather information and subsequently to produce a decision about what the car needs to do further. The modern computers are very powerful and can process large amounts of data at once at a speed that is far faster than human capabilities.

Now, consider the fact that most of the vehicles on road are manned, and their movement is unpredictable sometimes. For deploying AVs safely on the public roads, various tasks need to be analyzed. This includes observing as well as predicting the motion and trajectories of surrounding vehicles, navigating the AV safely to the destination, while taking all of the above into account.

Driving a vehicle manually can be a difficult task, as drivers have to keep in mind many things, like other vehicles around them, pedestrians, traffic lights, etc., and sometimes keeping track of all these things can become overwhelming and can lead to accidents. This is where smart vehicles and AVs come in, to reduce such risks. Cameras and sensors all around the car collect lots of real-time information and these can then be used by a carefully tailored computer program to detect objects of concern, follow on the desired path, and so on. Now, this approach could not have been possible even a few years back, due to the lack of computational power and the expense involved. But now the revolutionary change/development of the AI/ML sector in the industry has enabled us to carry on with these approaches, which are proving to be better and better by each passing day.

The main objective of this work is as follows:

- Carefully read and understand the data provided by Lyft.
- Train on the dataset by applying deep learning models.
- Optimize the loss function for training and validation dataset.
- Detect objects of concern in the video feed.

9.2 Related work

Now, this section includes the review of the other existing datasets used for training AVs from a classical state-of-the-art self-driving pipeline as shown below (Fig. 9.1):

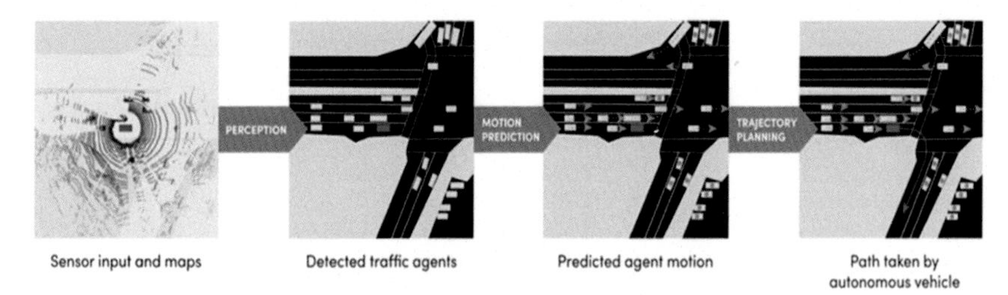

Figure 9.1 A classical SOTA (state-of-the-art) self-driving pipeline.

From left to right, the image shows the raw data taken from a sensor device called LiDAR, combining it with the data coming from the cameras. We can see the image after rendering produces a radar-like image where we can spot the position of the surrounding agents. The next image shows the data after the AV has processed the raw data, and this is basically what the AV sees. The blue object is the AV and the white ones are other agents. The third image shows the predicted motion of the surrounding agents. This is shown by collecting enough data on the other agents so as to accurately show their future movement direction. And at the end the AV plans its own trajectory to avoid all the agents and maintain the route toward its destination.

9.2.1 Perception datasets

Table 9.1 lists a few of the other perception datasets, these datasets are made by varying organizations but with a common main objective to supervise and estimate the 3D position of the peripheral agents. These have worked with similar problems related to object detection and semantic segmentation [7–10].

The KITTI dataset [2] focused on computer vision and self-driving vehicle-related problems. The GPS/IMU, LiDAR, front facing stereo cameras of KITTI have recoded about 6 hours of training data and 50 scenes. This work includes 3D bounding boxes with levels such as cars, trucks, and pedestrians. The Waymo Open Dataset [4] and nuScenes [11] are similar in size and structure. Both provide labeled 3D bounding boxes. The Oxford RobotCardataset [3] focuses on localization and mapping rather than 3D object detection and semantic segmentation.

Table 9.1 A comparison among various up to date AV datasets. In this chapter the Lyft Prediction Dataset is used.

Name	Size	Scenes	Map	Annotations	Task
KITTI [2]	6 h	50	None	3D bounding boxes	Perception
Oxford Robot Car [3]	1000 km	100 +	None	3D bounding boxes	Perception
Waymo Open Dataset [4]	10 h	1000	None	3D bounding boxes	Perception
ApolloScape Scene Parsing [5]	2 h	NA	None	3D bounding boxes	Perception
Argoverse 3D Tracking v1.1 [6]	1 h	113	Lane center lines, lane connectivity	3D bounding boxes	Perception
Lyft Perception Dataset	2.5 h	336	Rasterized road geometry	3D bounding boxes	Perception

9.3 Dataset distribution

The dataset is taken from the "Lyft 3D Object Detection for autonomous Vehicles" Kaggle dataset. The dataset is split into train and test set. The train set contains center_x, center_y, center_z, width, length, height, yaw, and class_name.

center_x, center_y, and center_z are the world coordinates of the center of the 3D bounding volume.

width, length, height are the dimensions of the volume.

Yaw is the angle of the volume around the z axis (where y is forward/back, x is left/right, and z is up/down—making "yaw" the direction the front of the vehicle/bounding box is pointing at while on the ground).

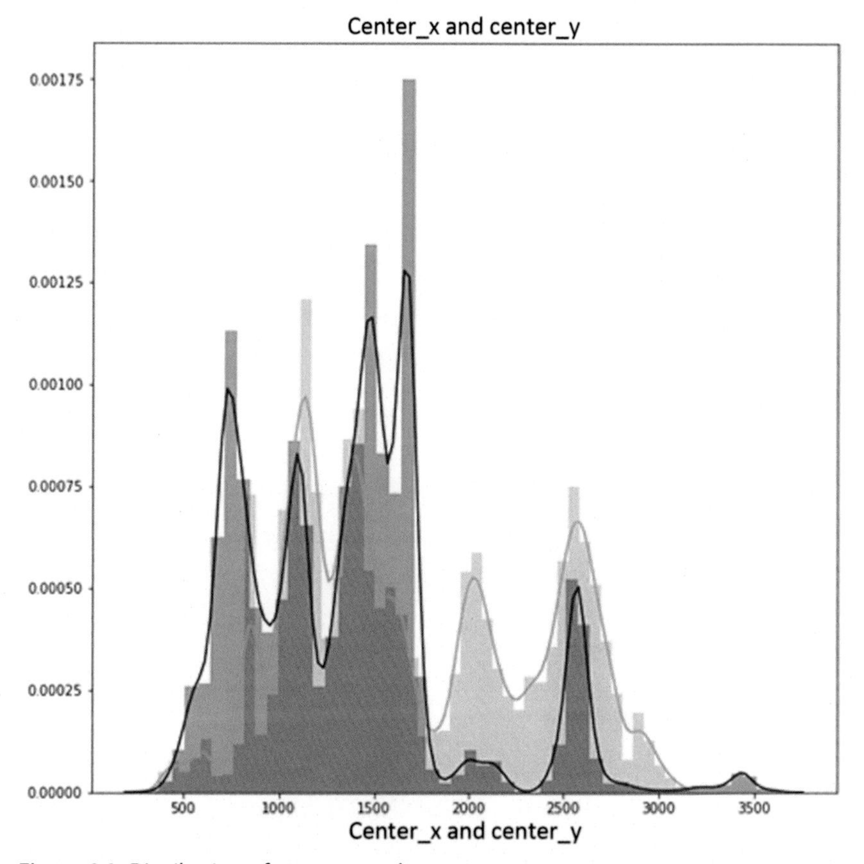

Figure 9.2 Distribution of center_x and center_y.

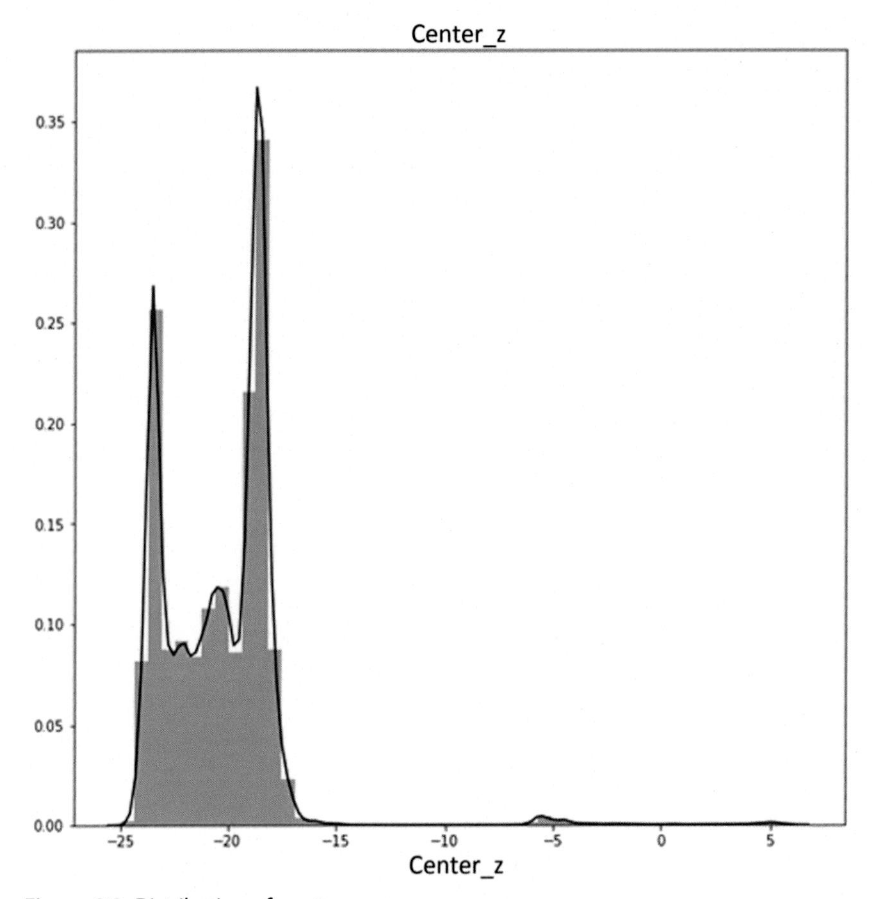

Figure 9.3 Distribution of center_z.

Class_name is the type of object contained by the bounding volume. The dataset distribution is given in Figs. 9.2–9.8.

9.4 Methodology

Four deep learning models U-Net, You Only Look Once (YOLO), VoxelNet to train the model. At first, the training dataset is fed to the network. After training, the models are evaluated.

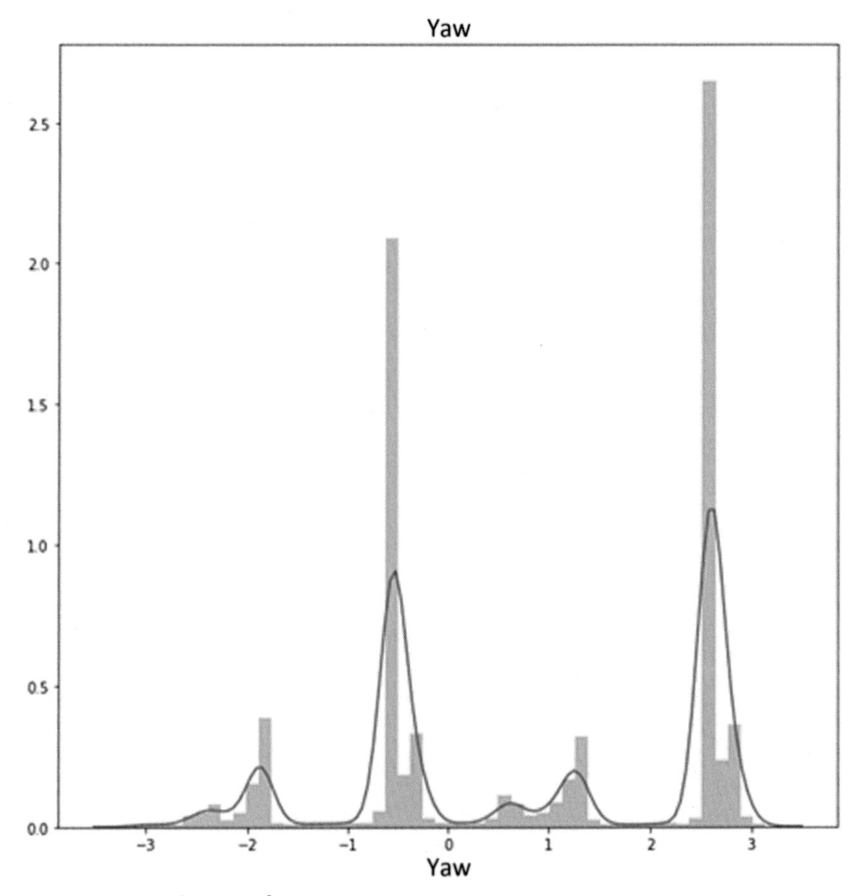

Figure 9.4 Distribution of yaw.

9.4.1 Models

UNET

In Fig. 9.9, we can see a classical UNet, where we can clearly see why the model is called so. Now if we take the left part of the "U," we can see the Convolution blocks + Relu activations and Max Pooling layers. To explain the working properly, we take an example of an input image which is of size $572 \times 572 \times 1$. We can now see how the convolutions

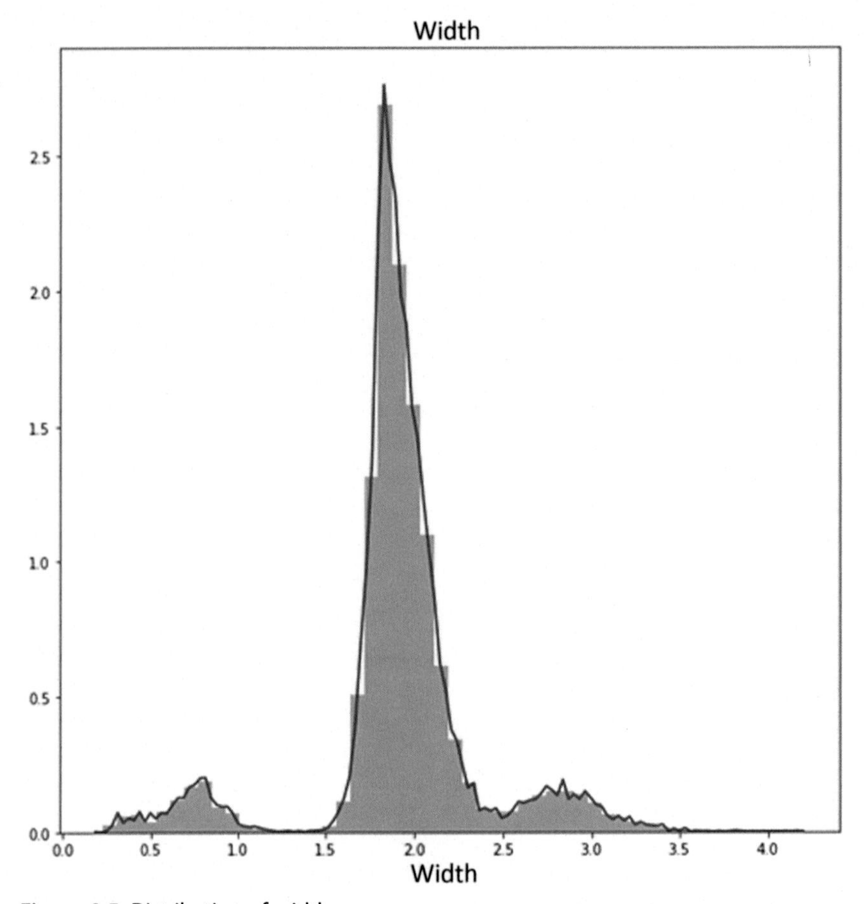

Figure 9.5 Distribution of width.

downscale the image by reducing it from 572 × 572 to 28 × 28 × 1024 at the bottom of the model.

Now coming to the right side of the "U," we see upscaling. The model reduces depth and increases the height and width of the image. If we take the bottom-up approach then we can see the image goes from 28 × 28 × 1024 to 56 × 56 × 1536 (first upsampling) to 102 × 102 × 256 (convolution to reduce depth) to 100 × 100 × 256 to 200 × 200 × 384 ...... 388 × 388 × 2.

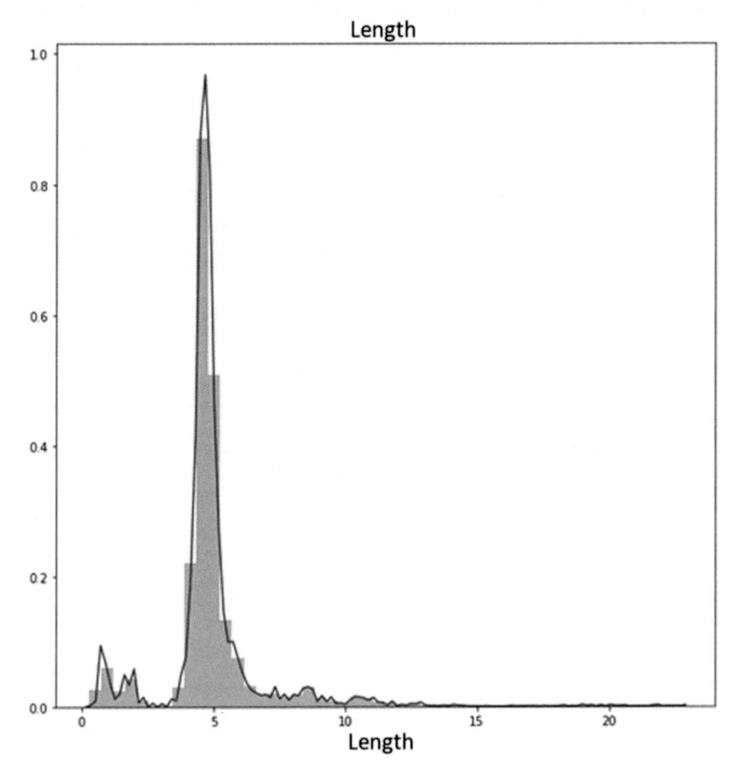

Figure 9.6 Distribution of length.

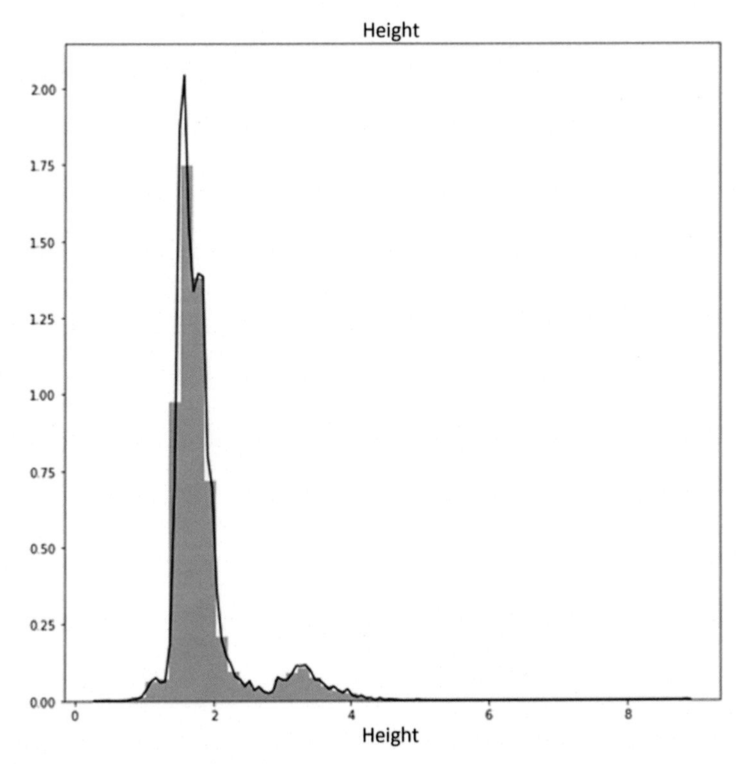

Figure 9.7 Distribution of height.

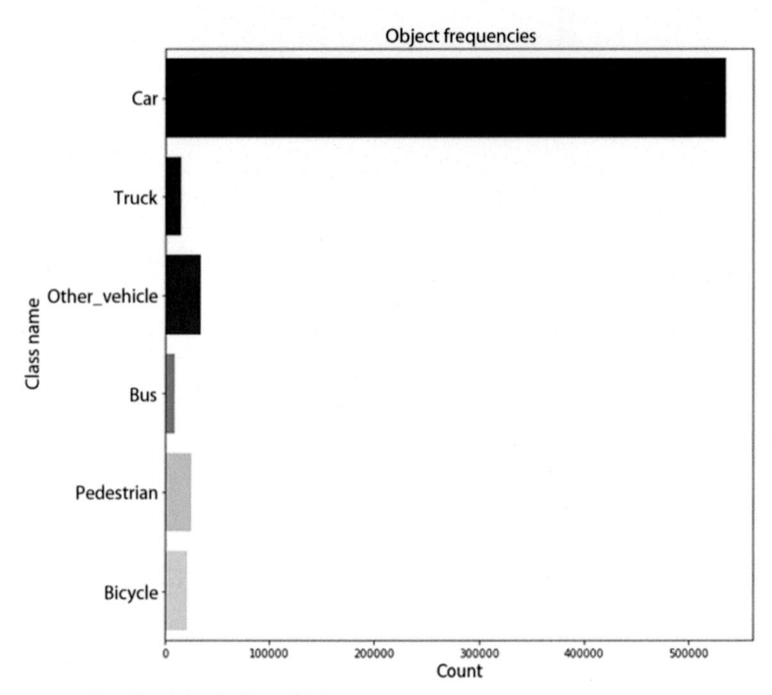

Figure 9.8 Distribution of class objects.

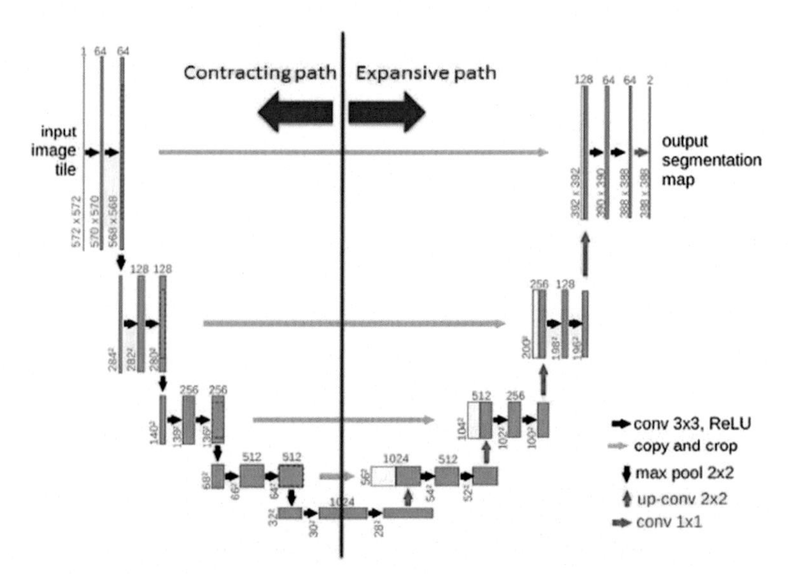

Figure 9.9 UNET architecture.

YOLO:

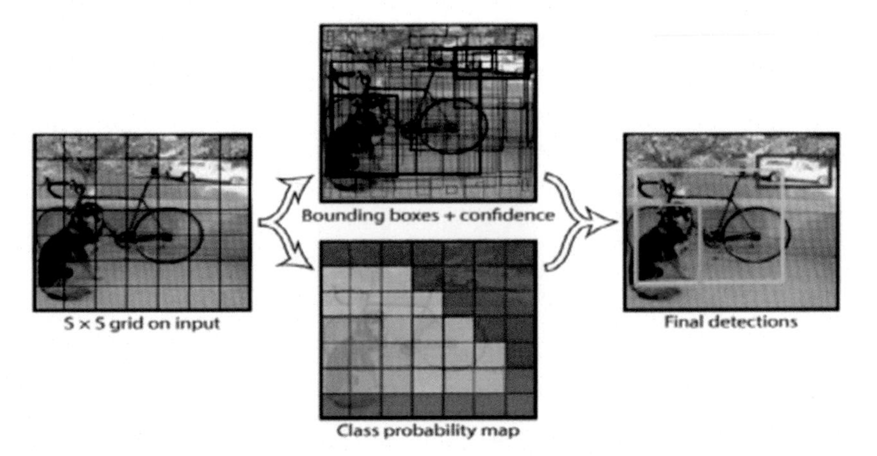

YOLOV3 is somewhat different than the other models here with similar motive (region proposal classification networks) that perform detection on multiple regions of the image and end up performing prediction multiple times on the same image but on different regions. However, YOLO is different, it resembles a fully convolutional neural network in which the image is passed only once through the entire network and the output is the prediction. Thus the design is cacophonic with the input image in m × m grid and for every grid there is the generation of a pair of bounding boxes and sophistication chances for those bounding boxes. Note that the bounding box is possible to be larger than the grid itself.

YOLO architecture:

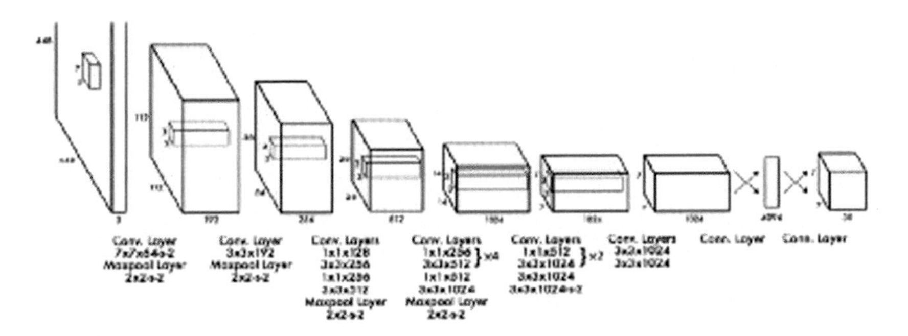

Our network has 24 convolutional layers followed by two absolutely connected layers. Rather than the beginning modules utilized by GoogLeNet, we tend to merely use one × one reduction layers followed by three × three convolutional layers.

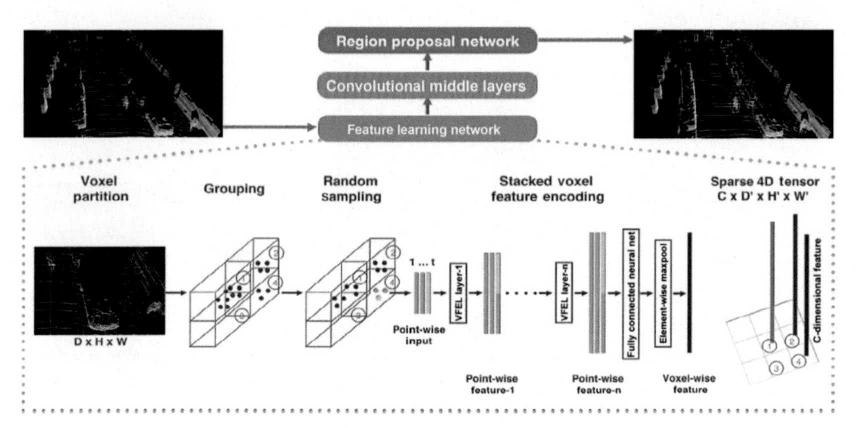

Figure 9.10 VoxelNet architecture.

Fast YOLO uses a neural network which has lesser convolutional layers (nine rather than 24) and fewer filters in those layers. Apart from the dimensions of the network, all coaching and testing parameters are constant between YOLO and quick YOLO.

VoxelNet:

This is an end-to-end connected network which combines feature extraction along with bounding box prediction. And to our convenience, it works directly on 3D point cloud data. We can see in Fig. 9.10 how the network generates 3D bounding boxes from the point cloud.

First, the entire 3D space is equally divided into voxels. Then the points are grouped according to the voxel they belong to.

The first layer of the architecture is actually an encoding layer which transforms a set of points within each voxel into a feature representation; 4D tensor. This layer is called "Voxel Feature Encoding layer."

After this, a 3D convolution is applied to the result to aggregate voxel-wise features. The output of this convolution middle layer is taken as the input for the Region Proposal Network (RPN) layer. The RPN produces a probability score map and a regression map. The RPN is shown below. The loss is the sum of the classification loss and the regression loss (Fig. 9.11).

Evaluation matrix:

A Confidence Score is calculated as evaluation standard. This Confidence Score shows the probability of the image being detected correctly by the algorithm and is given in percentages. The scores are taken on the mean average precision at different IoU, i.e., Intersection over

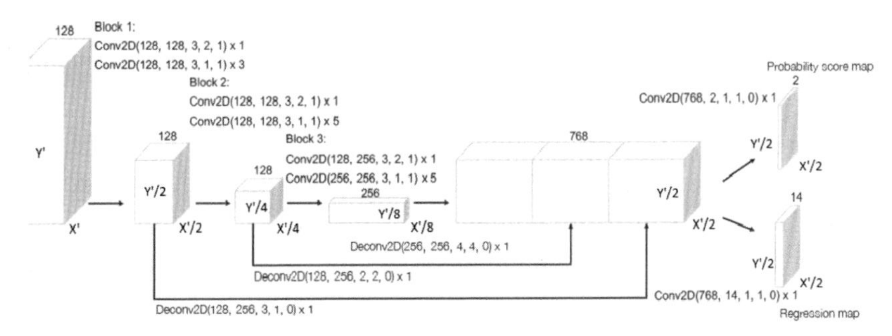

Figure 9.11 3D convolutional neural network.

Union threshold. For this work, the thresholds are set at 0.7, 0.75, 0.8, i.e., if the confidence of the detected object is over 70%, 75%, and 80%, respectively, then only the label will be taken for evaluation.

IoU is measured by the magnitude of overlap between the bounding boxes of two objects. The formula for IoU is given below:

$$\text{Intersection over Union} = \frac{\text{Area of Overlap}}{\text{Area of Union}} \tag{9.1}$$

Eq. (9.1) can be expressed as:

$$\text{IoU} = \frac{X \cap Y}{X \cup Y} \tag{9.2}$$

Precision is calculated at the threshold value depending on the True Positives (TP), False Positives (FP), and False Negatives (FN), which are the result of comparing the predicted object to the ground truth object. A true positive is counted when the ground truth bounding box is matched with the prediction bounding box. A false positive is counted when a predicted object had no association with the ground truth object. A false negative is counted when a ground truth object had no association predicted object. The confidence score is given as the mean over all the precision scores for all thresholds. The average precision is given as:

$$\text{Confidence} = \text{Average Precision} = \frac{1}{|\mathit{Thresholds}|}\sum_{T}\frac{TP}{TP + FP + FN} \tag{9.3}$$

9.5 Result

For qualitative results, successful and unsuccessful examples of detections were examined for the better understanding of model's performances. Figs. 9.12 and 9.13 shows the examples of detections where most of the area within bounding box subsumes the ground truth.

Figure 9.12 View from front camera.

Figure 9.13 View from back camera.

Though multiple ground truth bounding boxes for an object can be predicted theoretically. In the results the given bounding boxes were predicted by the algorithm. This scenario is considered as a successful detection as the box is correctly localized and this information is fed to the car controller, so that the car can take a decision for the next time step. Figs. 9.12–9.17 show the objects detected with the front, front_left, front_right, back, back_left, and back_right camera.

Figure 9.14 View from left front camera.

Figure 9.15 View from right front camera.

For quantitative results, mean Average Precision score was calculated. The Intersection over Union (IoU) metric was calculated, which takes the overlap between the ground truth box and predicted box, divided by the total area cover by both the ground truth box and predicted box. If this overlap is greater than the threshold value (for our case, we took 0.55), the class is predicted as true positive. The confidence score was calculated

Figure 9.16 View from back left camera.

Figure 9.17 View from back right camera.

Table 9.2 Confidence score calculated for traffic agent detection and classification.

Model	Object							
	Car	Truck	Pedestrian	Bicycle	Animal	Bus	Motorcycle	Emergency vehicle
YOLO-v3	98.73	84.39	96.76	82.88	78.90	69.99	77.52	67.98
VoxelNet	89.07	83.76	88.93	72.89	59.34	81.92	72.34	63.67
UNet	86.85	85.35	90.48	79.98	75.65	77.77	75.36	65.56

for each of the peripheral objects (car, truck, pedestrian, bicycle, animal, bus, motorcycle, emergency vehicle) for each deep network (YOLO-v3, VoxelNet, UNet), as shown in Table 9.2.

9.6 Conclusions

In this chapter, a comparative analysis among different deep networks has been provided where each network predicts a three-dimensional bounding box with a confidence score for the detected class for the images taken during autonomous driving. The data on which we have worked is provided by Lyft through a competition in Kaggle. It includes images from different cameras put on the car with pictures from many different angles; we also have LiDAR data which helped a lot in determining the 3D space around the AV. The data was well labeled and different aspects were clearly differentiated. On this data, we have used three different models for comparison: YOLO-v3, VoxelNet, and UNet. These are some very popular architecture for object detection purposes and hence we have used them to evaluate how good they actually are. Until now, various multimodal datasets are published. In the datasets most of the datasets are taken from the cameras and LiDAR. Our chapter has mainly focused on the RGB images and LiDAR data. After training and running the models for prediction, there are some things that became clear to us. Firstly, YOLO-v3 is the fastest performing architecture. If the priority is speed along with decent accuracy then YOLO-v3 is a good choice. Secondly, UNet was the slowest model, the reason being that it is based on a classical architecture with back propagation which consumes a lot of time, but the performance is not bad, it is rather good. And lastly, VoxelNet lands somewhat in the middle ground between YOLO and

UNet in terms of both speed and accuracy. Any of the three models can be used for real-world application but they all require some fine-tuning to perform better at the particular task given.

References

[1] N. Djuric, V. Radosavljevic, H. Cui, T. Nguyen, F.-C. Chou, T.-H. Lin, N. Singh, and J. Schneider, Uncertainty-aware short-term motion prediction of traffic actors for autonomous driving, IEEEXplore, 2020.
[2] A. Geiger, P. Lenz, C. Stiller, R. Urtasun, Vision meets robotics: the KITTI dataset, Int. J. Robot. Res. (2013).
[3] J. Gao, C. Sun, H. Zhao, Y. Shen, D. Anguelov, C. Li, and C. Schmid, Vectornet: encoding hd maps and agent dynamics from vectorized representation. Int. Conf. on Computer Vision and Pattern Recognition (CVPR), 2020.
[4] P. Sun, H. Kretzschmar, X. Dotiwalla, A. Chouard, V. Patnaik, P. Tsui, J. Guo, Y. Zhou, Y. Chai, B. Caine, V. Vasudevan, W. Han, J. Ngiam, H. Zhao, A. Timofeev, S. Ettinger, M. Krivokon, A. Gao, A. Joshi, Y. Zhang, J. Shlens, Z. Chen, and D. Anguelov, Scalability in perception for autonomous driving: Waymo open dataset, 2019.
[5] P. Wang, X. Huang, X. Cheng, D. Zhou, Q. Geng, and R. Yang, The apolloscape open dataset for autonomous driving and its application, Transactions on Pattern Analysis and Machine Intelligence (TPAMI), 2019.
[6] M. Chang, J. Lambert, P. Sangkloy, J. Singh, S. Bak, A. Hartnett, D. Wang, P. Carr, S. Lucey, D. Ramanan, and J. Hays, Argoverse: 3d tracking and forecasting with rich maps. Int. Conf. on Computer Vision and Pattern Recognition (CVPR), 2019.
[7] C.R. Qi, W. Liu, C. Wu, H. Su, and L.J. Guibas, Frustum pointnets for 3d object detection from RGB-D data. Int. Conf. on Computer Vision and Pattern Recognition (CVPR), 2018.
[8] Y. Zhou and O. Tuzel, Voxelnet: end-to-end learning for point cloud based 3d object detection, Int. Conf. on Computer Vision and Pattern Recognition (CVPR), 2018.
[9] M. Liang, B. Yang, Y. Chen, R. Hu, and R. Urtasun, Multitask multi-sensor fusion for 3d object detection, Int. Conf. on Computer Vision and Pattern Recognition, 2019.
[10] A.H. Lang, S. Vora, H. Caesar, L. Zhou, J. Yang, and O. Beijbom, Pointpillars: fast encoders for object detection from point clouds, Int. Conf. on Computer Vision and Pattern Recognition (CVPR), 2018.
[11] H. Caesar, V. Bankiti, A.H. Lang, S. Vora, V.E. Liong, Q. Xu, A. Krishnan, Y. Pan, G. Baldan, and O. Beijbom, nuScenes: a multimodal dataset for autonomous driving. arXiv preprint arXiv:1903.11027, 2019.

CHAPTER TEN

Recent trends in pedestrian detection for robotic vision using deep learning techniques

Sarthak Mishra and Suraiya Jabin
Department of Computer Science, Faculty of Natural Sciences, Jamia Millia Islamia, New Delhi, India

10.1 Introduction

A crowd comprises groups of individuals in an open or closed space. Behavior of the individuals in a crowd as a single entity and with each other defines the collective behavior of the crowd. Monitoring crowd through CCTV cameras has been a tedious task in analyzing crowd behavior and identifying anomalous behavior that may lead to harm. A number of challenges exist in crowd behavior analysis, one of which is detecting pedestrians with absolute accuracy in the high density crowded scenes, individually as well as collectively.

Robotic vision is a branch of computer vision that specializes in evolving human-aware navigation (HAN), autonomous driving, smart camera-based surveillance, etc. Deep learning techniques are now being fully used to develop vision-based robots for these purposes. The development of smart homes and cities has been constantly on the rise with the development of surveillance systems based on Deep Neural Networks.

A Deep Neural Network is designed by keeping in mind the problem domain, a feature extractor extracts the relevant features from the data, and the classifier classifies the data into an appropriate category. With the boom of Computer Vision in 2012, Convolutional Neural Networks (CNNs) have since been used to solve some of the complicated computer vision problems which appeared impossible two decades ago. This has given rise to deep learning as a new tool for solving the problems of automated crowd analysis and monitoring. Deep Neural Networks have tackled some of the major challenges in the past few years, like classification of images,

Artificial Intelligence for Future Generation Robotics.
DOI: https://doi.org/10.1016/B978-0-323-85498-6.00008-3

Figure 10.1 Overview of a crowd monitoring system.

localization of classified objects in images, object recognition, semantic, and instance segmentation. The devised deep learning techniques have also been able to achieve state-of-the art results on these robotic vision problems. Fig. 10.1 shows an overview of a crowd monitoring system.

The unprecedented growth of the population has demanded the immediate need of advanced surveillance and crowd monitoring systems that will provide unequivocal contributions to the field of computer vision. Advancement of deep learning, with efficient applications, tools, and techniques being developed frequently, has eased the task to make computer vision comparable to human vision, now not a distant reality. Deep learning models have become more robust and accurate in automating the process of pedestrian detection (PD), tracking, and reidentification in high-density occluded crowded scenes.

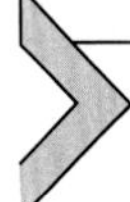

10.2 Datasets and artificial intelligence enabled platforms

A number of datasets are publicly available for PD, tracking, and identification. Here we list some of the widely used benchmark datasets in recent years.

- CityPersons [1]: contains 5k images with more than 35k annotations. It has on average seven pedestrians per image.
- NightOwls [2]: contains more than 279K annotations from three different countries. It is a night-time dataset.
- Caltech [3]: contains 10 hours of video from a vehicle driving in an urban scene having more than 42K training and 4K testing images.
- ETH [4]: contains three video sequences with more than 1.8K images.

- KAIST [5]: contains more than 95K images with 100K + annotations and 1182 pedestrian instances.
- INRIA [6]: contains 614 positive and 1218 negative images for training and 288 for testing.
- PETS [7]: contains video sequences from multiple cameras (eight).
- UMN [8]: contains 11 video sequences with three different scenes with more than 7K frames.

Here are some of the artificial intelligence (AI)-enabled platforms that offer surveillance and other detection capabilities.

- Ella from ICRealtime [9]: uses google cloud to process CCTV footage from any camera. It is a real-time open source platform that integrates deep learning tools for detection, recognition, and can be integrated with any number of cameras.
- AnyConnect [10]: provides smarter surveillance tools for CCTV footage using IP cameras. It fuses deep learning, robotic vision, and sensor information over a wireless network for bodycams and security cameras.
- Icetana [11]: offers surveillance through video anomaly detection by identifying abnormal behavior. It is a real-time monitoring platform that automatically learns normal patterns and recognizes anomaly.
- Hikvision [12]: offers facial recognition and vehicle identification management tools. Offers security with the help of multiple tools under the safe city program to monitor surveillance footage, alarm systems, and so on.

10.3 AI-based robotic vision

Robotic vision systems are being developed for HAN, human machine interaction, unmanned aerial vehicle (UAV), and autonomous driving. Such systems need an efficient vision system for object detection and localization. Deep learning techniques have improved the capabilities of robotic vision applications, such as automated surveillance, autonomous driving, drones, smart cities, etc. Such systems are highly complex and can't be easily developed with limited resources, but the advantages gained by developing such systems have a crucial role in improving the way of life for humans.

Kyrkou [13] designed a system to achieve a good trade-off between accuracy and speed for a deep learning-based PD system using smart cameras. The proposed system uses separable convolutions and integrates connections across layers to improve representational capability. The authors also suggest a new loss function to improve localization of detection. Mandal et al. [14] designed a real-time robotic vision system for multiobject tracking and detection in complex scenes using RGB and depth information. The spatiotemporal object representation combines a color model with the texture classifier that is based on the Bayesian framework for particle filtering.

Detection of pedestrians from cameras at varying altitudes is a complex issue. UAVs or drones are now being designed with robotic vision system for better trajectory planning. Jiang et al. [15] proposes a unique technique for PD to be employed in UAVs for efficient PD. The system is trained with Adaptive Boosting and cascade classifiers, based on Local Binary Patterns, with Meanshift algorithm for detection. Aguilar et al. [16] designed an extension of Jiang et al.'s work [15] where Saliency Maps algorithm is used instead of Meanshift. The dataset provided by the authors is suitable for use in UAV systems. The images are captured from surveillance cameras at different angles as well as altitudes.

For developing a mobile robot (or Mobot), robotic vision is one of the most important aspects. It is important for the Mobot to be aware of obstacles and which of them are people. An efficient mobile robot is supposed to have a robust vision system that can detect and localize pedestrians as well as plan and predict the trajectory of motion. Hence PD becomes an important aspect of robotic vision in terms of HAN. Mateus et al. [17] addressed the problem of HAN for a robot in a social aspect. This author proposed a deep learning-based person tracking system, where a cascade of aggregate channel features (ACF) detector with a deep convolutional neural network (DCNN) is used to achieve fast and accurate PD. Human-aware constraints associated with robots' motion were reformulated using a mixture of asymmetric Gaussian functions, for determining the cost functions related to each constraint. Aguilar et al. [18] applied the real-time deep learning techniques on human-aware robot navigation. The CNN is combined with Aggregate ACF detector that generates region proposals to save computational cost. The generated proposals are then classified by CNN. The suggested pedestrian detector is then integrated with the navigation system of a robot for evaluation. Riberio et al. [19] proposed a novel 3DOF pedestrian trajectory prediction method for autonomous mobile service robots. Range-finder sensors were used to learn and predict 3DOF pose trajectories. T-Pose-LSTM was trained to learn context-dependent human tasks and was able to predict precisely both position and orientation. The proposed approach can perform on-the-fly prediction in real time.

10.4 Applications of robotic vision toward pedestrian detection

In this section some of the major applications of PD using robotic vision are described. Fig. 10.2 shows these applications.

10.4.1 Smart homes and cities

In the smart home, town or cities, robotic vision is gaining pace in order to accommodate the growing demands of fully automated systems. These systems are supposed to detect and localize any object or person and can make decisions based on those detections. For example, a person enters a room, the system should be able to identify if it's one of the residents or an intruder and can raise the alarm. In smart towns or cities, detecting a pedestrian can be a first step toward a robust surveillance system that can identify any wanted person or detect any suspicious activity. Garcia et al. [20] presented the basic idea of using an IP camera to automate the process of person detection in smart homes, smart towns, and smart cities. The authors have used MIDGAR architecture which is a popular IOT platform and integrated it with a computer vision module to analyze images from the IP cameras. The real-time monitoring system developed can be used to define actions owing to the presence of a person in an area. Ibrahim et al. [21] designs a framework using Deep Convolutional Neural Network for detecting pedestrians in an urban scene, and also mapping slums and transport modes among other things. The Urban-I

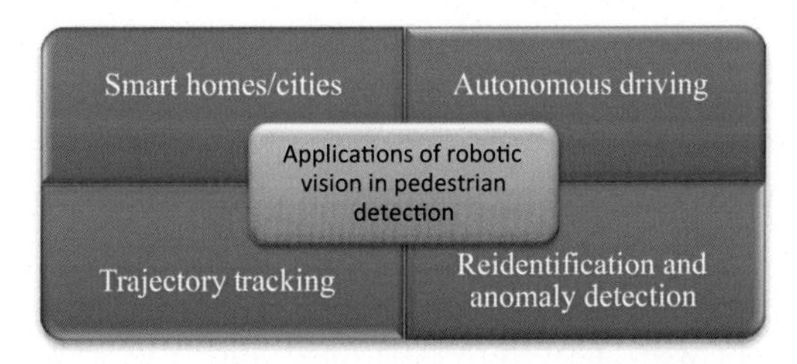

Figure 10.2 Applications of robotic vision towards pedestrian detection.

architecture comprises three parts. One of which is used for identifying pedestrians using the CNN and Single Shot Detector technique. The framework is useful in smart cities as it serves multiple purposes. Othman et al. [22] combined the IOT system and computer vision module like Garcia et al. [20] to detect people. Using the PIR sensor on a Raspberry 3 card, motion is detected in an area and an image is captured. The computer vision module is then used to detect the position of the pedestrian in that image. The resultant image is sent to a smartphone along with a notification. The proposed system is useful in maintaining security in a smart home. For detecting pedestrians at distance and to improve the resolution of input images, Dinakaran et al. [23] made use of Generative Adversarial Networks (GANs) in combination with a cascaded Single Shot Detector. GAN is helpful in image reconstruction to discriminate features of objects present in the image. This discriminative feature of the proposed system is helpful in differentiating vehicles and pedestrians from a street surveillance camera to make it suitable for smart cities applications. Lwowski et al. [24] proposes a fast and reliable regional detection system integrated with deep convolutional networks for real-time PD. This study has its application in robotics, surveillance, and driving. The proposed study used rg-chromaticity to extract relevant red, green, blue (RGB) channels features and a further sliding window algorithm was utilized for the detection of feature points for classification purposes.

10.4.2 Autonomous driving

Designing an autonomous vehicle or mobile robots crucially requires a robust and accurate PD system. Khalifa et al. [25] propose a dynamic approach to detect moving pedestrians using a moving camera placed on top of a car's windshield. This framework models the motion of the background using several consecutive frames then differentiates between the background and foreground, then checks for the presence of pedestrians in the scene. The main idea by Wang [26] is to develop a system of PD based on panoramic vision of the vehicle. Center and Scale Prediction, which is a two-stage process of feature extraction and detection, has been designed. Resnet-101 is used as base model which is used for feature extraction using multiple channels. In the detection stage, small-sized convolutions are used to predict the center, scale, and offsets. Pop et al. [27] proposes PD along with activity recognition and time estimation to cross the road for driver assistance systems in order to increase road safety. PD is done using RetinaNet, which is based on ResNet50, using the JAAD dataset for training that is annotated to derive the pedestrian actions before crossing the streets. The actions are classified into four categories: pedestrian is preparing to cross the street (PPC), pedestrian is crossing the street (PC), pedestrian is about to cross the street (PAC), and pedestrian intention is ambiguous (PA).

10.4.3 Tracking

For the purpose of crowd surveillance, it is important to detect pedestrians in a crowded scene, it is also important to track the pedestrians in a scene. The trajectory of pedestrian's movement is crucial in understanding the flow of the crowd, analysis of crowd behavior, and detection of anomalous behavior in crowd. Mateus et al. [17] proposes a PD and tracking mechanism using multicamera sensors. Detection is done by combining ACF detectors along with Deep Convolutional Neural Network. HAN for a robot in a social context is handled using Gaussian as the cost function for different constraints. In Chen et al. [28] a pedestrian detector is developed using a faster region based convolutional neural networks (R-CNN) and then a tracker is designed using a target matching mechanism by utilizing a color histogram in combination with scale invariant feature transformed into a Fully Convolutional Network to extract pedestrian information. It is also shown to reduce background noise in a much more efficient manner and works well in occluded conditions. Zhong et al. [29] presents a novel technique for estimating the trajectory of pedestrians' motion in 3D space rather than 2D space. A stereo camera is used to detect and estimate a human pose using the Hourglass Network. For predicting trajectory, SocialGAN is extended from 2D to 3D. The detection of the CNN can be improved by enhancing the representational ability. Yu et al. [30] proposes fusing extracted features such as color, texture, and semantics using different channels of the CNN by emphasizing the correlations among them via learning mechanisms. Yang et al. [31] proposes a pedestrian trajectory extraction technique using a single camera that doesn't acquire any blind space. A Deep CNN in combination with Kalman filter and Hungarian algorithm to detect pedestrian heads are used to obtain the coordinates of the trajectory points using a novel height estimation technique.

10.4.4 Reidentification

Reidentifying pedestrians is a challenging task. It involves tracking a pedestrian across multiple cameras/scenes as well distinguishing different people at the same time. A number of factors come into play like similarity in appearances, varying camera angles, pedestrian pose, and so on. Fu et al. [32] proposes a two-stream network based on ResNet50 that uses spatial segmentation and identifies global as well as local features. One of the two branches in the network learns global spatial features using adaptive average pooling, whereas the other branch learns the local features using horizontal average pooling to depict the pose of the pedestrians. Qu et al. [33] proposes a Deep Convolutional Neural Network that identifies differences in the local features between the two input images. The DCNN extracts feature vectors from the input images then utilize Euclidean distance to calculate the similarity. For training, the focal loss is used to handle the class imbalance problem in the dataset, thus improving the recognition accuracy of the overall system.

10.4.5 Anomaly detection

Anomaly in terms of surveillance refers to any suspicious activity that is different from the activities occurring in the scene. A dual channel CNN [34] enables identification of suspicious crowd behavior by integrating motion and scene-related information of the crowd extracted from input video frames. This highlights the significance of raw frames along with feature descriptors in identifying an anomaly in crowded scenes. Fully Convolutional Neural Networks (FCN) have also been used recently for detecting abnormal crowd behaviors. These networks can be used in combination with temporal data to detect and localize abnormalities in crowded scenes [35]. Transfer Learning (pretrained CNN) can be adapted to an FCN (by removing the Fully Connected Layers). This FCN-based structure extracts distinctive features of input regions and verifies them with the help of a special-design classifier. Ganokratanaa et al. [36] attempted to detect anomalies and localize them using a deep spatiotemporal translation network based on GAN. Using normal activities framed as a training set, the authors train a system to generate dense optical flow as temporal features of normal events. For application, the input videos are fed into the system and any event that was not in the training set is categorized as anomaly.

10.5 Major challenges in pedestrian detection

In this section, we list the most pertinent issues faced in the area of state-of-the-art PD and remedies suggested by various researchers. Fig. 10.3 shows the major challenges faced by PD.

10.5.1 Illumination conditions

Illumination conditions play the most crucial role in accurate PD. Several factors can contribute to bad lightening conditions, such as bad weather, nighttime, or insufficient light in an indoor scene.

Researchers [37] designed a multispectral PD system which consists of two subnetworks designed using region-based FCN, by combining color and thermal image information. The model fuses the color and thermal images, and nonmaximum suppression is used to fuse the detection results of the two subnetworks. Researchers [38] proposed a multispectral Deep Convolutional Neural Network that combines the visible and infrared images using a two-branch network which is merged together using three different fusion techniques at different convolutional stages. Color and thermal images and their correlation can also be used as a measure to detect pedestrians in nighttime videos. This illumination measure [39] can be used as a facilitator in the detection process or can be used to design a CNN for the same. CNN fusion architectures have also proved useful in

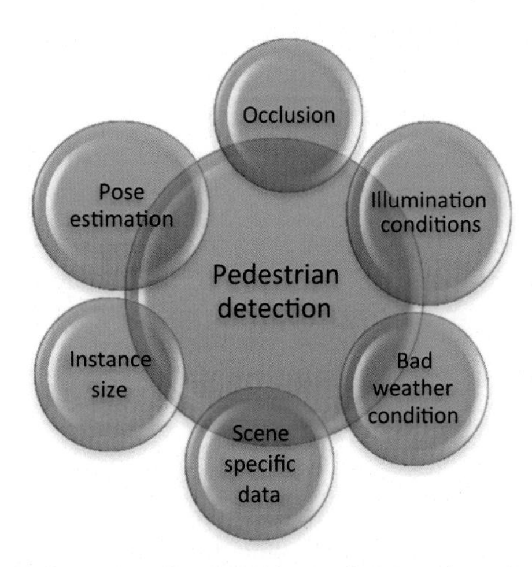

Figure 10.3 Major challenges in the faced by pedestrian detection.

detecting pedestrian in images with illumination problem. Vanilla architectures have been used to obtain and combine the detection results of the fusion architectures. In outdoor scenarios, weather plays an important role as visibility is totally dependent on it. Tumas et al. [40] focused on detecting pedestrians in adverse weather conditions. A new dataset named ZUT captured during severe weather conditions (cloudy, rainy, foggy, etc.) was used to train a system trained using You Only Look Once (YOLO) v3 detector. The detector performs well in varying lighting conditions and under motion blur as well.

10.5.2 Instance size

Position of the camera that captures a scene can have positive as well as negative effects on how a system detects pedestrians in that scene. For example, pedestrians that are closer to the camera can be easily identified but the small instances can even be missed by a human eye. When Deep Neural Networks process these scenes, there is a possibility that they may miss the small instances.

A Scale Aware Fast R-CNN synthesizes two subnetworks into a single framework [41], and detects pedestrians in outdoor scenes where different pedestrian instances possess different spatial scales. The framework takes the location of people (proposal) and the feature map of the scene as inputs, and outputs a prediction scores and a boundary for each input proposals. The framework in [38] uses a Feature Pyramid Network that is shown to improve the detection results for different scales of instances (especially, small-scale instances). A two-stream CNN [42], replicating the human brain, captures spatial and optical flow of the input video sequences for detecting pedestrian using deep motion features and deep still image features. Zhang et al. [43] focused on the detection of small-scale pedestrians using an asymmetrical CNN by considering the shape of the human body as rectangular. This facilitates to generate rectangular proposals that capture the complex features of a human body. A three-stage framework for the CNN deploys coarse-to-fine features to gently dismiss the false detections.

10.5.3 Occlusion

Occlusion has proved to be one of the major hindrances in PD. It becomes quite difficult to detect a pedestrian who is being blocked by an object or another pedestrian.

Mask R-CNN can be used to detect and segment faces in obscured images [44]. This is one way of handling occlusion and increasing robustness of face detection mechanism. GANs can be another intuitive way to handle occlusion. A generative and adversarial architecture integrated with attribute classification can be trained using transfer learning and a discriminator [45]. The generator generates the new image and the discriminator tries to authenticate it, depending on how well it is trained. This is a cool way of reconstructing occlusion-free images of a person which is credible in possible scenarios. Extensive Deep Part detectors are another way of handling occlusion. These detectors cover all the scales of different parts of human body and choose significant parts for occlusion handling [46]. Each part detector can be individually trained for different body parts using a quality training data or transfer learning. The output of each part detector can then be combined using weighted scheme or a direct score. A semantic head detection system [47] in parallel with the body detection system is used to handle occlusion. The head is one of the most crucial parts while detecting pedestrians. Predictions are inferred from the body detector and are used to manually label the semantic head regions for the training process. A Deep Neural Network has taken under consideration the semantic correlation among different attributes of the body parts [48]. The system uses element wise multiplication layer that avails the feature maps to extract different body feature representations from the input. Fei et al. [49] uses context information from the extracted features from the network to handle occlusion. A pixel-level context embedding technique to access multiple regions within the same image for context information using multibranched layers in CNN with varying input channels and an instance-level context prediction framework identifies distinguishing features among different pedestrians in a group.

10.5.4 Scene specific data

One of the factors in the limitation of the performance of the detection models is unavailability of the scene-specific dataset. A model can be trained on different types of generic dataset, but when it comes to real-world applications, it suffers due to scarcity of scene-specific datasets.

In an online learning technique to detect pedestrian in a scene-specific environment, Zhang et al. [50] designed a model on an augmented reality dataset which addresses the issue of unavailability of scene-specific data. Synthetic data is used to simulate scenes of need (which is also available publicly in [51]). The same model is gradually improved over the time as a specific scene changes. Cyget et al. [52] explored the importance of data augmentation toward improving the robustness of PD systems, evaluated the robustness of different available augmentation techniques, and then proposed a novel technique that uses Style-transfer, combining the input image and stylized version of the image in patches to maintain balance. Inspired from Patched Gaussian Augmentation, this work adds patches from the styled image to the original image at the same location to improve detection. Tumas et al. [40] introduced a new dataset, ZUT (Zachodniopomorski Uniwersytet Technologiczny), captured during severe weather conditions (cloudy, rainy, foggy, etc.) from four EU countries (Denmark, Germany, Poland, and Lithuania). It was collected using a 16 bit thermal camera at 30fps and is fine-grained annotated in nine different classes: pedestrian, occluded, body parts, cyclist, motorcyclist, scooterist, unknowns, baby carriage, pets and animals.

10.6 Advanced AI algorithms for robotic vision

With the advent of CNNs, deep learning approaches have been refined steadily for building a more robust pedestrian detector. Based on different challenges posed in the real-world situations, different types of Deep Neural Networks have been designed to improve the robotic vision. Over the years the systems have evolved significantly from preprocessing to detection to localization. Initially CNNs advanced to region-based CNN that gave rise to fast and faster R-CNNs which changed process of detection with the use of Region Proposal Networks (RPNs). Researchers further exploited these RPNs to extract customized spatial and temporal features to enhance trajectory plotting in a scene based on detection. Further, multistream Nets have been deployed to multiple instances as well as different body parts to handle occlusion in a scene. This has proved ingenious as pedestrians can also be recognized with their body parts even when they are partially visible.

Multistream Nets have also been able to manipulate color and thermal version of an image simultaneously to incorporate illumination changes. Multispectral or thermal images have significantly been used in recent works to enhance detection at nighttime or during bad weather. These images are incorporated with the normal version to observe changes and find differences. For data, researchers have also started generating synthetic data for scene-specific studies. As mentioned above, a number of datasets are publicly available and benchmarked for use but designing a real-world detection system still poses a serious challenge. GANs have been used frequently to generate data and simulate specific scenes. These help the system learn different scenarios and also be prepared for what it hasn't yet seen (Table 10.1).

Table 10.1 Summary of the papers reviewed. The deep learning technique they used, challenges they tackled (if any), dataset on which they performed validation, and their performance.

References	Deep learning method used	Challenges	Dataset	Result metrics	Performance
[25]	SVM	Illumination, nighttime	CVC-14	F1-Score	94.30 for Day time 87.24 for Night time
[37]	Region-Based Fully CNN (Feature Pyramid Network)	Instance size, illumination	KAIST	FPPI	0.34
[38]	CNN + Focal Loss	Illumination	KAIST	Miss Rate (%)	27.60
[34]	CNN	–	UMN	AUC (%)	Greater than 97
			PETS		Greater than 92
[41]	Fast R-CNN	Instance size	Caltech	Miss Rate (%)	9
			INRIA		8
			ETH		34
[39]	Faster R-CNN	Illumination	KAIST	Miss Rate (%)	15.73
[35]	FCN (Transfer learning with AlexNet)	–	UCSD Ped2	EER (%)	11
			Subway		16
[46]	CNN (Transfer Learning on ImageNet)	Occlusion	Caltech	Miss Rate (%)	11
			KITTI	Average Precision (%)	74
[2]	CNN	Illumination	KITTI	Accuracy (%)	70
[42]	CNN	Motion features	Caltech	Miss Rate (%)	16.7
			DaimlerMono		35.2
[17]	CNN	–	INRIA	Unspecified	Unspecified
			MBOT		
[47]	Head Body Alignment Network	Occlusion	CityPersons	Log Average Miss Rate (%)	11.26 on reasonable and 39.54 on heavy Occlusion

(Continued)

Table 10.1 (Continued)

References	Deep learning method used	Challenges	Dataset	Result metrics	Performance
[44]	Mask R-CNN	Occlusion	LFW	AUC (%)	77
[45]	GAN (Transfer Learning with Resnet and VGG)	Occlusion	RAP AiC	Mean Accuracy (%)	81 90
[48]	CNN	Occlusion, instance size	RAP PETA	Average Accuracy (%)	92.23 91.70
[43]	Asymmetrical multi-stage CNN	Instance size	Caltech CityPersons	Miss Rate (%)	7.32 13.98
[50]	Faster R-CNN	Scene specific	TownCenter Atrium PETS 2009	Average Precision (%)	34 12.5 0.9
[30]	Faster R-CNN with VGG16	Feature representation	Caltech ETH	Miss Rate (%)	16.5 19.65
[31]	CNN with Kalman Filter	—	Custom data	Pixel error	5.07
[26]	CSP with ResNet-101	—	CityPersons	Log Average Miss Rate (%)	9.3
[28]	Faster R-CNN	Occlusion	OTB-50	Graphical	-
[32]	Two stream ResNet-50	Pose estimation	Market-1501 DukeMTMC CHUK03	Mean Average Precision (%)	90.78 84.82 71.67
[49]	CNN with ResNet-50	Occlusion	Caltech CityPersons	Miss Rate (%)	44.48 11.4
[29]	Hourglass Network	Pose estimation	Custom data	Error Reduction (%)	47
[36]	SpatioTemporal Translation Network using GAN	—	UCSD UMN CUHK	AUC (%)	98.5 99.96 87.9

[40]	YOLOv3	Bad weather, nighttime, scene specific	Custom Data	Mean Average Precision	89.1
[33]	CNN + Euclidean Distance + Focal Loss	–	CUHK03	Accuracy (%)	Rank1 – 76 Rank10 – 95.6 Rank20 – 99.5
[52]	Faster R-CNN and CSP	Data augmentation	CityPersons EuroCity NightOwls	ECE	0.1429 0.1569 0.331
[27]	RetinaNet using ResNet-50 and LSTM	Activity recognition	JAAD	Mean Average Precision (%)	56.05

10.7 Discussion

A number of challenges have influenced PD but tackling them has been a giant step toward creating a lucrative robotic vision system for implementation in various fields. What recent years have achieved in this domain is incomparable to what we could only dream of a decade ago. The evolution of technology in terms of processing power has made it easier for the researchers across the globe to design sophisticated deep learning techniques that can even perform better than human experts. Starting from detecting and localizing objects and moving toward localizing anomalies in CCTV footage has indeed made it a significant achievement.

Fig. 10.4 shows timeline of evolution of deep learning techniques for PD. Evolution of self-driven cars has been possible due to accurate detection by Deep Neural Networks (DNN) and how well they process the input. Anomaly detection is another very important application of pedestrian detectors that has been built to identify any outlier activity in a scene. Some of the discussed work has tackled this issue and given promising results in terms of response time and recognition of the anomaly. Suspicious activities are context related and thus systems need to be designed to be able to process those contexts. Tracking and reidentification are a part of anomaly detection task where a person of interest (PoI) is tracked across the scene and should be reidentifiable across multiple cameras in a building (or place) in consecutive scenes. A number of factors come into play: human pose, camera angle, similarity index, and so on. All these factors need to be handled with precision to make those systems reliable and robust.

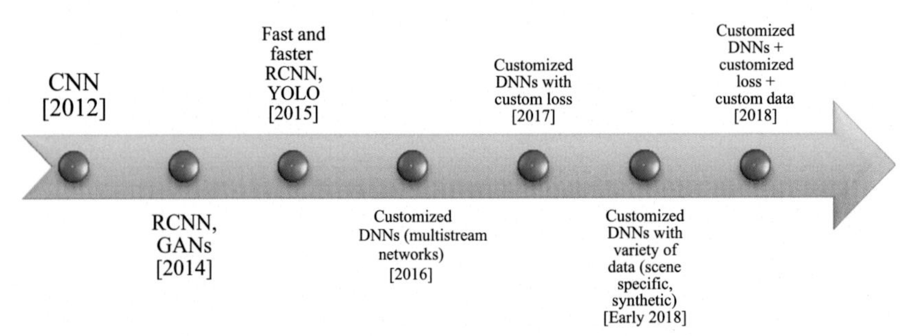

Figure 10.4 Evolution of deep learning methods for pedestrian detection over the years. *CNN*, Convolutional Neural Network; *GAN*, Generative Adversarial Network.

Mobile robots (or Mobots) are a significant achievement in the field of robotic vision. Detecting the obstacles (or persons) and predicting their trajectory in order to plan navigation comes under HAN. Several high-end systems have shown promising results in the field with a chance for improvement as well. These systems can be further optimized to design automated delivery systems, rescue operations, healthcare system, UAVs, and so on. All these robotic vision systems developed use deep learning algorithms in some way and have been optimized for better results as well.

Although the detectors are performing quite well on the available benchmark datasets, it will be interesting to see how they perform in real-world situations. None of the systems have been fully developed for use as a practical application tool in public places to replace humans. At some point or other they need humans in the loop. There are two significant reasons: none of the benchmark or available dataset can actually simulate real-world scenario, and there will always be something new (unique) that system does not know how to handle. This issue has gained a lot of attention in recent years. With the introduction of GAN, it is possible to create synthetic data according to one's needs and create scene-specific detectors for real-world use.

Another reason is occlusion, no machine has been able to fully handle occlusion and there are justifications behind it. How much occlusion can human vision handle? We as a human learn to recognize things even if we see a part of it but that doesn't happen overnight. It takes years of experience of practicing intuitions and familiarity. Researchers have tried designing part detectors, that can detect different parts of a body and infer a detection based on that, but still they have not been entirely accurate at times. It can be considered as a start toward a more robust and reliable solution for providing a better tool for future use.

10.8 Conclusions

A decade ago, pedestrian detection posed numerous challenges to the pioneers in computer vision. Identifying people in a video and differentiating them from other objects was crucial in solving the problem of crowd monitoring. Object detection through CNNs paved the way for detecting people in crowded scenes. Robotic vision has also seen a boom with these techniques. Smart homes, mobots, and autonomous vehicles are all being upgraded as a result of this development in robotic vision. Preprocessing the input data has also been crucial in difficult detection scenarios, which implies that data and its handling is as important as the tool it is being used with. Apart from that, several application-based systems have also been developed to automate the surveillance process and monitor crowded places. In the long run, deep learning has proven to be the most effective tool in evolving robotic vision systems and provide a useful baseline for other systems to work upon. In the coming years, we will be able to see these systems perform more robustly under occlusion and bad lighting/weather conditions, which is the growing need.

References

[1] S. Zhang, R. Benenson, and B. Schiele, Citypersons: a diverse dataset for pedestrian detection, in: Proceedings of the IEEE Conference on Computer Vision and Pattern Recognition, 2017, pp. 3213–3221.
[2] L. Neumann, M. Karg, S. Zhang, C. Scharfenberger, E. Piegert, S. Mistr, et al., NightOwls: a pedestrians at night dataset, in: Asian Conference on Computer Vision, Springer, Cham, December 2018, pp. 691–705.
[3] P. Dollár, C. Wojek, B. Schiele, and P. Perona, Pedestrian detection: a benchmark, in: 2009 IEEE Conference on Computer Vision and Pattern Recognition, IEEE, June 2009, pp. 304–311.
[4] N.B. Weidmann, J.K. Rød, L.E. Cederman, Representing ethnic groups in space: a new dataset, J. Peace Res. 47 (4) (2010) 491–499.
[5] Y. Choi, N. Kim, S. Hwang, K. Park, J.S. Yoon, K. An, et al., KAIST multi-spectral day/night data set for autonomous and assisted driving, IEEE Trans. Intell. Transp. Syst. 19 (3) (2018) 934–948.
[6] G. Overett, L. Petersson, N. Brewer, L. Andersson, and N. Pettersson, A new pedestrian dataset for supervised learning, in: 2008 IEEE Intelligent Vehicles Symposium, IEEE, June 2008, pp. 373–378.
[7] J. Ferryman and A. Shahrokni, Pets2009: dataset and challenge, in: 2009 Twelfth IEEE International Workshop on Performance Evaluation of Tracking and Surveillance, IEEE, December 2009, pp. 1–6.
[8] R. Mehran, A. Oyama, and M. Shah, Abnormal crowd behavior detection using social force model, in: 2009 IEEE Conference on Computer Vision and Pattern Recognition, IEEE, June 2009, pp. 935–942.
[9] L. Sun, Z. Yan, S.M. Mellado, M. Hanheide, and T. Duckett, 3DOF pedestrian trajectory prediction learned from long-term autonomous mobile robot deployment data, in: 2018 IEEE International Conference on Robotics and Automation (ICRA), IEEE, May 2018, pp. 1–7.
[10] https://smartella.com
[11] https://anyconnect.com/
[12] https://icetana.com/
[13] C. Kyrkou, YOLOpeds: efficient real-time single-shot pedestrian detection for smart camera applications, IET Comput. Vis. 14 (7) (2020) 417–425.
[14] S. Mandal, S. Biswas, V.E. Balas, R.N. Shaw, A. Ghosh, Motion prediction for autonomous vehicles from Lyft Dataset using deep learning, 2020 IEEE 5th International Conference on Computing Communication and Automation (ICCCA), 30–31 Oct. 2020, pp. 768–773, Available from https://doi.org/10.1109/ICCCA49541.2020.9250790.

[15] M. Jiang, D. Wang, T. Qiu, Multi-person detecting and tracking based on RGB-D sensor for a robot vision system, Int. J. Embed. Syst. 9 (1) (2017) 54–60.
[16] W.G. Aguilar, M.A. Luna, J.F. Moya, V. Abad, H. Parra, and H. Ruiz, Pedestrian detection for UAVs using cascade classifiers with meanshift, in: 2017 IEEE 11th International Conference on Semantic Computing (ICSC), IEEE, January 2017, pp. 509–514.
[17] A. Mateus, D. Ribeiro, P. Miraldo, J.C. Nascimento, Efficient and robust pedestrian detection using deep learning for human-aware navigation, Robot. Auton. Syst. 113 (2019) 23–37.
[18] W.G. Aguilar, M.A. Luna, J.F. Moya, V. Abad, H. Ruiz, H. Parra, et al., Pedestrian detection for UAVs using cascade classifiers and saliency maps, International Work-Conference on Artificial Neural Networks, Springer, Cham, 2017, pp. 563–574. June.
[19] D. Ribeiro, A. Mateus, P. Miraldo, and J.C. Nascimento, A real-time deep learning pedestrian detector for robot navigation, in: 2017 IEEE international conference on autonomous robot systems and competitions (ICARSC), IEEE, April 2017, pp. 165–171.
[20] C.G. García, D. Meana-Llorián, B.C.P. G-Bustelo, J.M.C. Lovelle, N. Garcia-Fernandez, Midgar: detection of people through computer vision in the Internet of Things scenarios to improve the security in Smart Cities, Smart Towns, and Smart Homes, Future Gener. Comput. Syst. 76 (2017) 301–313.
[21] M.R. Ibrahim, J. Haworth, T. Cheng, URBAN-i: from urban scenes to mapping slums, transport modes, and pedestrians in cities using deep learning and computer vision, Environ. Plan. B Urban Anal. City Sci. (2019). 2399808319846517.
[22] N.A. Othman and I. Aydin, A new IoT combined body detection of people by using computer vision for security application, in: 2017 9th International Conference on Computational Intelligence and Communication Networks (CICN), IEEE, September 2017, pp. 108–112.
[23] R.K. Dinakaran, P. Easom, A. Bouridane, L. Zhang, R. Jiang, F. Mehboob, et al., Deep learning based pedestrian detection at distance in smart cities, in: Proceedings of SAI Intelligent Systems Conference, Springer, Cham, September 2019, pp. 588–593.
[24] J. Lwowski, P. Kolar, P. Benavidez, P. Rad, J.J. Prevost, and M. Jamshidi, Pedestrian detection system for smart communities using deep Convolutional Neural Networks, in: 2017 12th System of Systems Engineering Conference (SoSE), IEEE, June 2017, pp. 1–6.
[25] A.B. Khalifa, I. Alouani, M.A. Mahjoub, N.E.B. Amara, Pedestrian detection using a moving camera: a novel framework for foreground detection, Cognit. Syst. Res. 60 (2020) 77–96.
[26] W. Wang, Detection of panoramic vision pedestrian based on deep learning, Image Vis. Comput. 103 (2020) 103986.
[27] D.O. Pop, A. Rogozan, C. Chatelain, F. Nashashibi, A. Bensrhair, Multi-task deep learning for pedestrian detection, action recognition and time to cross prediction, IEEE Access 7 (2019) 149318–149327.
[28] K. Chen, X. Song, X. Zhai, B. Zhang, B. Hou, Y. Wang, An integrated deep learning framework for occluded pedestrian tracking, IEEE Access 7 (2019) 26060–26072.
[29] J. Zhong, H. Sun, W. Cao, Z. He, Pedestrian motion trajectory prediction with stereo-based 3D deep pose estimation and trajectory learning, IEEE Access 8 (2020) 23480–23486.
[30] P. Yu, Y. Zhao, J. Zhang, X. Xie, Pedestrian detection using multi-channel visual feature fusion by learning deep quality model, J. Vis. Commun. Image Represent. 63 (2019) 102579.
[31] L. Yang, G. Hu, Y. Song, G. Li, L. Xie, Intelligent video analysis: a Pedestrian trajectory extraction method for the whole indoor space without blind areas, Computer Vis. Image Underst. (2020). 102968.

[32] M. Fu, S. Sun, N. Chen, D. Wang, X. Tong, Deep fusion feature presentations for nonaligned person re-identification, IEEE Access 7 (2019) 73253–73261.
[33] W. Qu, Z. Xu, B. Luo, H. Feng, Z. Wan, Pedestrian re-identification monitoring system based on deep convolutional neural network, IEEE Access 8 (2020) 86162–86170.
[34] Y. Xu, L. Lu, Z. Xu, J. He, J. Zhou, C. Zhang, Dual-channel CNN for efficient abnormal behavior identification through crowd feature engineering, Mach. Vis. Appl. (2018) 1–14.
[35] M. Sabokrou, M. Fayyaz, M. Fathy, Z. Moayed, R. Klette, Deep-anomaly: fully convolutional neural network for fast anomaly detection in crowded scenes, Comput. Vis. Image Underst. 172 (2018) 88–97.
[36] T. Ganokratanaa, S. Aramvith, N. Sebe, Unsupervised anomaly detection and localization based on deep spatiotemporal translation network, IEEE Access 8 (2020) 50312–50329.
[37] L. Ding, Y. Wang, R. Laganière, D. Huang, S. Fu, Convolutional neural networks for multispectral pedestrian detection, Signal Process. Image Commun. 82 (2020) 115764.
[38] D. Pei, M. Jing, H. Liu, F. Sun, L. Jiang, A fast RetinaNet fusion framework for multi-spectral pedestrian detection, Infrared Phys. Technol. 105 (2020). 103178.
[39] C. Li, D. Song, R. Tong, M. Tang, Illumination-aware faster R-CNN for robust multispectral pedestrian detection, Pattern Recognit. 85 (2019) 161–171.
[40] P. Tumas, A. Nowosielski, A. Serackis, Pedestrian detection in severe weather conditions, IEEE Access 8 (2020) 62775–62784.
[41] J. Li, X. Liang, S. Shen, T. Xu, J. Feng, S. Yan, Scale-aware fast R-CNN for pedestrian detection, IEEE Trans. Multimed. 20 (4) (2018) 985–996.
[42] R. Yoshihashi, T.T. Trinh, R. Kawakami, S. You, M. Iida, T. Naemura, Pedestrian detection with motion features via two-stream ConvNets, IPSJ Trans. Comput. Vis. Appl. 10 (1) (2018) 12.
[43] S. Zhang, X. Yang, Y. Liu, C. Xu, Asymmetric multi-stage CNNs for small-scale pedestrian detection, Neurocomputing 409 (2020) 12–26.
[44] B. Kneis, Face detection for crowd analysis using deep convolutional neural networks, in: International Conference on Engineering Applications of Neural Networks, Springer, Cham, September 2018, pp. 71–80.
[45] F. Fulgeri, M. Fabbri, S. Alletto, S. Calderara, R. Cucchiara, Can adversarial networks hallucinate occluded people with a plausible aspect? Comput. Vis. Image Underst (2019).
[46] Y. Tian, P. Luo, X. Wang, and X. Tang, Deep learning strong parts for pedestrian detection, in: Proceedings of the IEEE international conference on computer vision, 2015, pp. 1904–1912.
[47] R. Lu, H. Ma, and Y. Wang, Semantic head enhanced pedestrian detection in a crowd, Neurocomputing, 2020.
[48] E. Yaghoubi, D. Borza, J. Neves, A. Kumar, and H. Proença, An Attention-Based Deep Learning Model for Multiple Pedestrian Attributes Recognition, 2020. arXiv preprint arXiv: 2004.01110.
[49] C. Fei, B. Liu, Z. Chen, N. Yu, Learning pixel-level and instance-level context-aware features for pedestrian detection in crowds, IEEE Access 7 (2019) 94944–94953.
[50] W. Zhang, K. Wang, Y. Liu, Y. Lu, F.Y. Wang, A parallel vision approach to scene-specific pedestrian detection, Neurocomputing 394 (2020) 114–126.
[51] http://openpv.cn/
[52] S. Cygert, A. Czyżewski, Toward robust pedestrian detection with data augmentation, IEEE Access 8 (2020) 136674–136683.

Further reading

Available at: <https://www.hikvision.com/en/solutions/solutions-by-industry/safe-city/>.

S. Kim, S. Kwak, B.C. Ko, Fast pedestrian detection in surveillance video based on soft target training of shallow random forest, IEEE Access 7 (2019) 12415–12426.

Index

Note: Page numbers followed by "*f*" and "*t*" refer to figures and tables, respectively.

F

G

H

I

K

L

M

N

O

P

R

S

Printed in the United States
by Baker & Taylor Publisher Services